SpringerBriefs in Astronomy

Series Editors

Wolfgang Hillebrandt, MPI füf Astrophysik, Garching, Germany
Martin Ratcliffe, Adjunct Lecturer, Wichita State University, Kansas, USA, and
Director, Professional Development, Sky-Skan, Inc.

For further volumes:
http://www.springer.com/series/10090

J. Woods Halley

How Likely is Extraterrestrial Life?

 Springer

Prof. J. Woods Halley
School of Physics and Astronomy
University of Minnesota
Church St SE 116
Minneapolis
MN 55455
USA
e-mail: woods@woods1.spa.umn.edu

ISSN 2191-9100   e-ISSN 2191-9119
ISBN 978-3-642-22753-0   e-ISBN 978-3-642-22754-7
DOI 10.1007/978-3-642-22754-7
Springer Heidelberg Dordrecht London New York

Library of Congress Control Number: 2011936145

Printed on acid-free paper

Springer is part of Springer Science+Business Media (www.springer.com)

Preface

This volume was produced for use in a series of undergraduate seminars that I have taught at the University of Minnesota for more than 10 years on the title subject. I began teaching the seminars, which continue and have variously been part of the undergraduate honors and freshman seminar programs at the university, for two reasons. I thought that the subject lent itself to a review of materials from a variety of disciplines, making it a good pedagogical vehicle for introducing undergraduates to a broad view of the fields in which they might later specialize, while learning some useful intellectual skills. Secondly, I was personally interested in deepening my own understanding of the title question.

Both motives seem to me to have been amply justified. Students in the seminars, who write a paper and give a talk on a topic related to the subject, show frequent indications of enthusiasm and maturation during the course, while my views of the subject question have significantly evolved. In this context, the mathematical level here has been kept to elementary algebra and trigonometry, except in the appendices, where a little calculus is occasionally used.

I started with the conference proceedings [1] edited by Zuckerman and Hart, which is still frequently cited here, and which, as I discovered over time, takes, on the whole, a quite 'negative' view of the entire question relative to other available writings. That is to say, many contributors to that volume believed that extraterrestrial life was unlikely to be found. Through the years, as I have read a great deal more, I have understood that other views are, in the present state of knowledge, at least as likely to be correct and I have tried to reflect that understanding in the presentation here. The result, I think, is a quite 'agnostic' account, which is not committed to one answer to the title question. As I discuss particularly in the last chapter, that uncommited perspective is somewhat unusual in published work on the subject, some of which is strongly devoted to one hypothesis or another.

I think that this 'agnosticism' is pedagogically important for at least two reasons: First it is, I think, an honest statement of our present understanding of this question. Secondly, the exercise of resisting the temptation to leap to a conclusion or strong hypothesis, which is particularly strong with regard to this intriguing question both for students and professionals, is a very useful one in the training of young minds. Occasionally, I have students practically begging me to tell them what I 'really think' is the answer to the title question. What I 'really think' is that the answer is not known, though one can constrain the list of possible answers by a careful survey of what is known, as I have tried to do here. The intellectual tension resulting from an attempt to assimilate this point of view, is, I think, very instructive. Individuals who might wish to use this monograph in a teaching context are invited to contact the author about teaching materials including power point lecture slides and homework exercises.

Among the colleagues with whom I have interacted on subjects related to this work, I particularly thank John Broadhurst, Robert Gehrz, Alexander Grossberg and Robert Pepin. Robert Pepin not only discussed many aspects with me, but spoke often to the seminar and read and commented on a significant portion of the manuscript. Graduate student Ivan Fedorov, who has been working with me on models of prebiotic evolution and the more than 100 students who have taken the seminar have also had a big influence on my thinking. However I am solely responsible for all the conclusions and views expressed here.

Reference

1. B. Zuckerman, M.H. Hart, *Extraterrestrials, Where Are They?*, 2nd edn. (Cambridge Press, Cambridge, 1995)

Minneapolis, 21 June 2011   J. Woods Halley

Contents

Chapter 1
Introduction

Human speculation concerning the existence of life recognisably like our own in other parts of the universe has a very long history. In ancient times, the space outside the earth was believed to be extensively inhabited by superhuman gods with traits very much like those of humans. During subsequent centuries, the geometrical structure of the universe gradually became clearer as a result of better observation interpreted by improved mathematical theory. In the seventeenth century, through the efforts of Copernicus, Kepler, Newton and many others, the nature of the solar system was established. In the nineteen twenties Hubble established the basic outlines of our present picture of the universe as an enormous rapidly expanding collection of galaxies. As this scientific picture evolved, it became more difficult to accomodate the existence of other human-like civilizations in it. The imagined family of man-like gods was replaced in most religious belief by a single deity whose physical nature and location was usually not specifically described. However a fascinated curiosity about the possibility of other human like beings persists. As recently as 1909, respected astronomer Lowell seriously advanced the idea that 'canals' on Mars were evidence of a form of extraterrestrial life there [1]. Subsequent observation has eliminated this hypothesis, but speculation about the existence of other civilizations farther afield continues. It is significantly encouraged by the enormous size and the very long times which are associated with current models of the universe. (The long history of speculation on the subject of extraterrestrial life has been particularly well reviewed by Michael Crowe [2] and by George Basalla [3].)

Astronomers can only observe stars whose light has had time during the age of the universe to reach earth. Stars which are close enough to satisfy this requirement are said to be part of the observable universe. In Chap. 2, we will review some aspects of what is known about the observable universe. It consists of about 10^{11} galaxies (Fig. 1.1) each containing about 10^{11} stars (Fig. 1.2) and is about 14 billion years old. Superficially it seems very likely that in such an old and spacious universe there would be many civilizations like our own. Significant (private) resources, are spent to support the activities of serious well trained scientists to look for them each year in the various Search for Extraterrestrial Intelligence (SETI) programs. In the US,

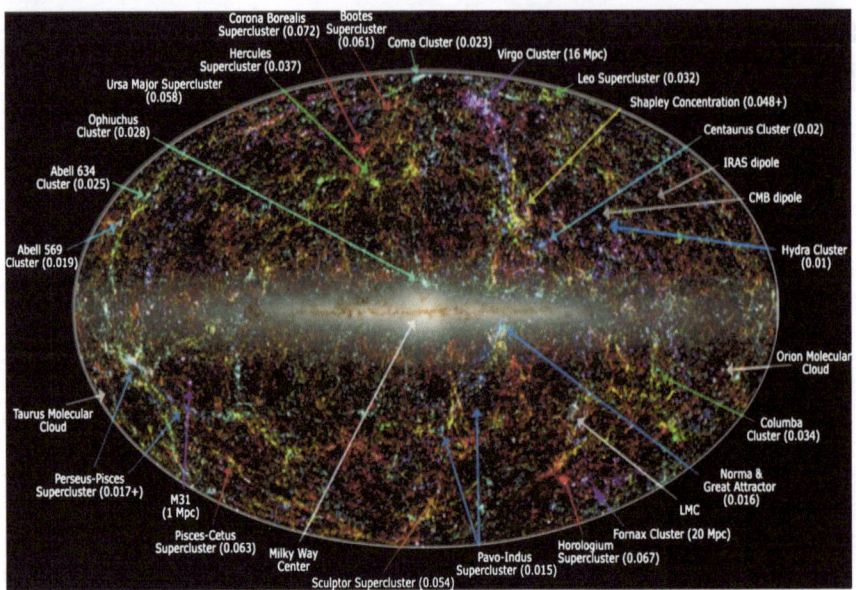

Fig. 1.1 A picture of the entire sky in the infrared showing observable clusters of galaxies. Estimates of the number of galaxies in the observable universe are of the order 10^{11} so the number of observable stars is of order $10^{11} \times 10^{11} = 10^{22}$. Image from Ref. [4]

Fig. 1.2 The spiral galaxy M74, NGC628 which is similar to our own. Such galaxies contain roughly 10^{11} stars. Image from Ref. [5]

the National Aeronautics and Space Administration (NASA) of the government has programs to search for evidence of life in the solar system but does not carry on SETI research seeking evidence of advanced civilizations.

One might think that a subject which arouses such deep human interest would have been the object of extensive theoretical study and that some definitive conclusions about, for example, the likelihood that SETI searches will succeed, would be available. This is not really the case. A great deal has been written, but very few definitive conclusions have emerged. For example, estimates of the number of advanced civilizations in our own galaxy range from one (ourselves, foregoing definitions of 'advanced' for the moment) to millions. Because it is fascinating to the public, extraterrestrial life has been a favorite subject of speculative fiction in books and films. Further, this fascination has led many untrained people to become involved in speculations which many believe. (There is some further discussion of such speculations in Chap. 5.) From the scientific point of view, no extraterrestrial civilizations or life have been observed, so one may say there are no data. Experimental probes of solar system planets funded by NASA have yielded no evidence for life of any kind and quite extensive privately funded radio searches for signatures of extraterrestrial intelligence have given entirely negative results.

One may question whether massive government funding and involvement of larger numbers of professional scientists at this time would improve the situation. As we will see, the subject immediately takes us to very deep and difficult questions concerning particularly the biochemistry associated with the origin of life and the nature of the processes which lead to the formation of planets and which control their climate. Nevertheless a great deal of reliable information bearing on the question of the likelihood of extraterrestrial life and civilization is available. Much of it will be reviewed here. This information does constrain the possibilities significantly compared to what might inferred superficially. My objectives in presenting this material are to summarize relevant known facts at the present (2009) and to point to the range of possiblities which they imply in as unbiased and scientific a manner as possible. The account which follows differs from some others in avoiding a strong hypothesis concerning the answer to the title question. It appears that, possibly because of the deep implications which it holds for the human condition, even highly qualified scientists have sometimes found such strong hypotheses irresistible. At the present, knowledge is insufficient to definitively choose between several very different possibilities (summarized in the last chapter as options 1 through 3.) and to hypothesize one of these and neglect the others cannot be justified. Nevertheless some information, particularly with regard to the occurence of extrasolar planets and the nature of heredity and the genome, is relatively new and has changed the perspective. Though we do not know precisely how likely we are to encounter civilizations or primitive life-like phenomena elsewhere in the universe, nor whether we could recognise them if we did, we can eliminate some of the scientific speculations which have been entertained even quite recently.

The book is divided into two parts with a concluding overview. In the first part, the question is approached in a manner which might be termed reductive or 'bottom up'. That is, we ask the question: What does existing scientific knowledge about physics, chemistry, climatology and biology tell us about the likelihood of extraterrestrial life and civilization? Answers to this question can be presented in an organised matter by working from the Drake equation, which is introduced below. The use of the Drake equation in this section is not intended to imply that it is an extremely useful predictive tool, but to organise the discussion. On the other hand, I do attempt to take the equation seriously and produce quantitative estimates for the factors appearing in it. It turns out to be quite possible to do this for the factors involving astronomy and planetary science, but attempts to estimate the probability for life to appear on a hospitable planet lead to huge uncertainties and concomitant uncertainty in the estimated number of expected life systems or civilizations per galaxy. However the way in which these uncertainties arise is very instructive and helps to focus attention on the key issues. Within the framework of the Drake equation we will discuss, in separate chapters, the frequency of appearance of suitable stars, the frequency of the appearance of suitable planets and the frequency with which complex life appears on suitable planets. (The definitions of some of the terms here will obviously need further consideration.)

In the second part, we consider a different inductive or 'top down' approach to the question. That is, we use the only data available to us which bears directly on the question, namely the fact that there is currently no credible scientific evidence for the existence of extraterrestrial biospheres or civilizations. As several authors have emphasized, this fact does have a significant bearing on the question. The account presented here is distinguished by some attempts to quantify the significance of the null results. This is a very common practise in other fields of science such as cosmology and high energy physics, where quantitative bounds are placed on the possible properties of hypothesized entities (such, for example, as the Higgs boson) which have not been observed, by using the parameters and properties of experiments which have failed to observe them. In the case of extraterrestrial life I will argue that existing data do permit some such bounds on the possibilities to be deduced. The larger intent of these attempts, however, is to encourage more such quantitative analysis of observations which fail to find extraterrestrial biospheres by various means. In Part II, I discuss the reasons for scientific rejection of UFO reports as evidence for the existence of extraterrestrial civilization, and the significance of the failure to observe colonization by extraterrestrials. Current and past radio and optical searches for extraterrestrial intelligence are reviewed and the significance of their failure to observe any signals over several decades of effort is discussed. The null results of efforts by NASA and other space agencies to find any form of life on other planets of our solar system are the subject of a penultimate chapter.

In the last chapter I summarize the possible answers to the title question in terms of three options, briefly termed 'life is dull', 'life is weird' and 'life is rare'. I conclude that the data reviewed leave each option a finite probability of describing the correct direction for answering the question. The human implications for policy, ethics and the human condition more generally of each of the three options are very different and these implications are briefly discussed.

1.1 The Drake Equation

We will focus on the question of how many biospheres and civilizations are likely to exist in a galaxy like our own, though similar considerations would apply to the larger question of the number to be expected in all observable galaxies. In the first part of the

book, we follow Drake [6] and attempt to estimate the expected number of civilizations or biospheres by considering separately the frequency of the elements required
for the existence of a biosphere or civilization. Because a biosphere is assumed
to be necessary for a civilization, the questions are closely related. Immediately
we encounter the questions of how to identify a 'biosphere' and how to define a
'civilization'. In either case how would we know when we had encountered or
observed one? For the present we provisionally suppose that a 'biosphere' consists
of a system of life-like autocatalytic nonequilibrium chemical processes occurring
on the surface of a planet and that a 'civilization'consists of a collection of individuals in such a biosphere, constructed of a carbon based chemistry much like our
own and processing energy and reproducing in ways similar to those found on earth.
We will later further suppose that this collection of individuals can produce complex
electromagnetic signals which are distinguishable from those produced by the nonbiological natural processes of which we are aware. Such a definition is not completely
unambiguous and it may be too restrictive but it will serve to begin the discussion.
It could be paraphrased roughly by saying that we are looking for something much
like us but not exactly. One imagines (as many writers have) differences like those
between Columbus and the native Americans in the fifteenth century. The differences
might very well be unimaginably greater and we will return to this particularly in
Chaps. 4 and 7.

With such a definition, one can enumerate the requirements for the existence of
a 'civilization' or a 'biosphere': There must be a star providing energy and a planet
on which the living system can live. Of course we did not say that the civilizations
or other biological systems had to live on a planet in our definition. However there is
no evidence that interstellar chemistry is sufficiently complex to generate life with a
chemistry like our own. Therefore, although the civilizations we seek currently might
not be living on a planet or planets, it seems very likely that they must have initially
evolved on planets. The star must be of an appropriate type, to be discussed later,
so that the requisite chemical elements needed for the chemistry are available and
the planet must be of an appropriate temperature and climate so that this chemistry
can occur in liquid water. (The uniqueness of water as a solvent will be discussed
in Chap. 3.) The star and the planets must exist long enough for life to evolve
on them and then the life must further evolve to sufficient complexity to evolve
complex electromagnetic signals, by which our scientists propose to detect it. Drake
summarized all these requirements in an equation:

$$N_{civ} = N_{gal} f_{star} f_{planet} f_{life} \qquad (1.1)$$

(This version of the equation is similar, and equivalent, to one given by Michael
Hart [7]. It is also equivalent to the original form of the equation as written by Drake.
The relationships between various forms of the Drake equation are explained in the
Appendix 1.1.) Here N_{civ} is the expected number of civilizations per galaxy and
N_{gal} is the average number of stars per galaxy. f_{star} is the probability that a star has
the needed properties, f_{planet} is the conditional probability that, having selected a
star, one finds a planetary system around the star which contains a planet with the

needed climate and chemistry and f_{life} is the probability that, given there is an earth like planet, life with the needed complexity has evolved on such a planet. Each of the probabilities can be further factored to take account of the various requirements. For example, f_{planet} can be regarded as the product of the probability that the star has a planetary system at all, times the probability that the planetary system contains a planet orbiting in a 'habitable zone' with an appropriate climate. If we are interested in the number of stars harboring planets with any kind of life at all (as in NASA searches) then an equation of the same form can also be used, with less restrictive conditions on the events for which the factor f_{life} gives the probability. If we are interested in the number of civilizations in the observable universe then at this level we need only substitute the total number of observable stars for the number of stars in a typical galaxy in the equation (There are roughly 10^{11} galaxies in the observable universe, with roughly 10^{11} stars per galaxy. See Figs. 1.1 and 1.2).

In the next chapters we will consider the factors in the Drake equation one at a time, starting with the 'astrophysical' factors $N_{gal} f_{star}$, proceeding to the planetary and meterological factor f_{planet} and finally discussing the biological factor f_{life}. However to get a feeling for what is involved, we note here that the number N_{gal} is quite well known (by the standards of this subject) to be of order 10^{11} or one hundred billion (Fig. 1.2).

It is this huge number which leads many to jump to the conclusion that N_{civ} must be large. As we will discuss in Chap. 2, f_{star} is approximately known and is of order 10^{-2}. Knowledge about f_{planet} is rapidly accumulating. It may be as large as 10^{-3}. The big unknown is f_{life}. Two points of view are found in the literature of the subject. One point of view, largely taken by physicists and enthusiasts of optimistic results for SETI searches, holds that f_{life} must be close to one because there is evidence that life appeared on earth very early in its history. If $f_{life} \approx 1$ and using the other estimates just cited, the Drake equation gives $N_{civ} \approx 10^6$ so there must be of the order of a million civilizations in each galaxy. The other point of view, likely to be taken by biologists more familiar with the complexities of the chemistry of life than their physicist colleagues, holds that f_{life} must be much smaller. Attempts to estimate f_{life} from detailed knowledge of the required biochemistry lead to numbers so small that the Drake equation leads to values of N_{civ} which are much less than 1! (This would simply mean that not every galaxy has a civilization). In this second scenario, the early appearance of life on earth might be explained by 'seeding' from elsewhere or by an event of very low probability. An attractive feature of the second scenario is that it would make it much easier than it is in the first scenario to understand why we have not observed any extraterrestrial civilizations, as we will discuss in Part II.

References

1. http://ssed.gsfc.nasa.gov/tharsis/canals.html
2. M. Crowe, *The Extraterrestrial Life Debate*, 1750–1900 (Dover Publications, New York, 1999)
3. G. Basalla, *Civilized Life in the Universe* (Oxford University Press, Oxford, 2006)
4. Courtesy of Dr. T. M. Jarrett (IPAC/Caltech) by permission

5. National Optical Astronomy Observatory/Association of Universities for Research in Astronomy/National Science Foundation (NOAO/AURA/NSF) by permission
6. F.D. Drake, Phys. Today **14**, 40 (1961)
7. M. Hart, in *Extraterrestrials, Where Are They?* ed. by B. Zuckerman, M. Hart, 2nd edn. (Cambridge University Press, Cambridge, 1995), p. 218

Part I
Bottom Up: What We Learn from Basic Science About the Likelihood of Extraterrestrial Life

In this section we trace what is known about the factors on the right hand side of the Drake equation. As we work from left to right, the scale, astronomical to planetary to molecular biology, gets smaller and the uncertainties in the estimates get larger. We begin in Chap. 2 with the astronomical factors $N_{gal} f_{star}$.

Chapter 2
Astrophysical Factors

$$N_{civ} = N_{gal} f_{star} f_{planet} f_{life}$$
$$\uparrow$$

To understand something of the elements which determine f_{star} we must review some aspects of the current understanding of the history of the universe. Evidence arising from observations of the rate at which stars and galaxies are receding from the earth indicates that, approximately 14 billion years ago (the exact time remains in some dispute) the universe as we observe it was confined to a very small region of space and has been expanding from that confined space ever since (the 'Big Bang') (More precisely, the space itself has been expanding, but this distinction need not concern us.) Some details of the very early history of that expansion remain in some dispute. However, there is strong evidence that, roughly one second after the expansion began, the initial material cooled sufficiently to leave mainly electromagnetic radiation, two kinds helium nuclear isotopes, two kinds of hydrogen nuclei (protons and deuterons) and electrons. At about 300,000 years after the initial explosion, the electrons combined with the nuclei to leave mainly electrically charge neutral atoms of helium and hydrogen as well as electromagnetic radiation.

The fact that the universe is expanding was first established in the 1920s, mainly by observations of Edwin Hubble of the Doppler shifts in the spectra of light emitted by stars in galaxies (see Appendix 2.1). The galaxies of stars outside the solar system were all found to be receding (redshifted spectra) from us with velocites v which increase with the distance r of the galaxies from us according to the relation v = Hr. H is called the Hubble constant. If the galaxies have been moving at a constant velocity then it is easy to see (Fig. 2.1) that this relation is consistent with convergence of all the galaxies at a common point at a time 1/H ago: Consider two galaxies 1 and 2 observed at distances r_1 and r_2 from us and moving at velocities v_1 and v_2. If the velocities obey the Hubble relation then $r_1 = (1/H)v_1$ and $r_2 = (1/H)v_2$ which are

J. W. Halley, *How Likely is Extraterrestrial Life?*, SpringerBriefs in Astronomy,
DOI: 10.1007/978-3-642-22754-7_2, © The Author(s) 2012

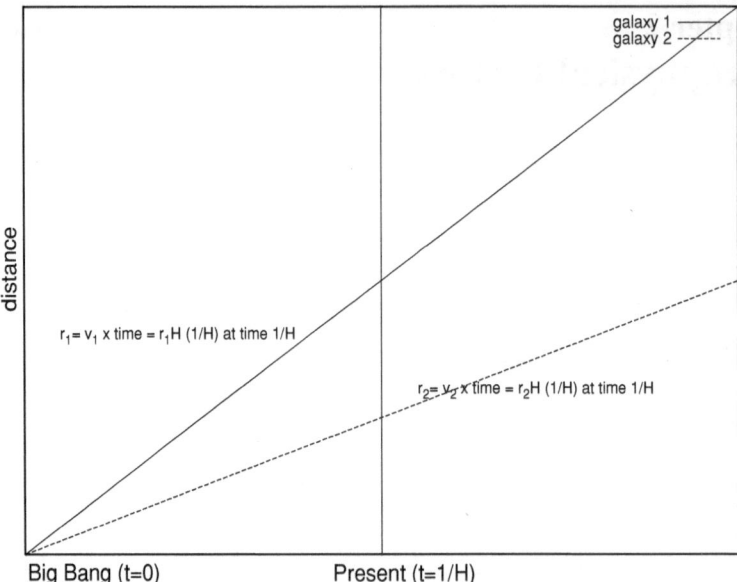

Fig. 2.1   Illustration of the relation of the Hubble constant to the age of the universe

equations describing two points moving at their respective velocities starting at a common point a time 1/H ago.

It has recently been discovered that the picture is complicated by indications that the galaxies are *accelerating* away from us, but the basic idea that the universe has been expanding from a common point for about 14 billion years is retained. There are two additional kinds of experimental evidence for this 'big bang' picture of the evolution of the universe: It predicts correctly the abundance of the helium and hydrogen as observed today and it accounts in considerable detail for the distribution of electromagnetic radiation observed in the universe.

For our purposes a very important point here is that in the scenario sketched so far, no carbon, oxygen or any of the other elements except hydrogen which are essential for life on earth have appeared. While it might be possible to imagine behavior of condensed hydrogen and helium which was complex enough to be characterized as 'life', nothing of that sort has ever been observed and we will suppose that a more complete collection of atoms from the periodic table is required for life. For example, a summary of those elements of the periodic table which are believed to be essential for human life appears in Table 2.1.

The rest of the atoms of the periodic table are known to have arisen from processes occuring in the natural 'nuclear furnaces' in the interiors of stars. This occurred in the following way. After the formation of the helium and hydrogen as described in the last paragraph, the resulting gas cloud gradually separated into clumps as a result of the mutual gravitational attraction that the atoms exerted on one another. These

Table 2.1 Periodic table, with elements required for human life underlined

1	2	3	4	5	6	7	8	9	10	11	12	13	14	15	16	17	18
1 **H** 1.01																	2 He 4.00
3 **Li** 6.94	4 **Be** 9.01											5 **B** 10.81	6 **C**	7 N	8 O	9 F	10 Ne 20.18
11 **Na**	12 **Mg**											13 **Al**	14 **Si**	15 **P**	16 **S**	17 Cl	18 Ar 39.95
19 **K**	20 **Ca**	21 **Sc** 44.96	22 **Ti**	23 **V**	24 **Cr**	25 **Mn**	26 **Fe**	27 **Co**	28 **Ni**	29 **Cu**	30 **Zn**	31 Ga 69.72	32 Ge	33 As	34 Se	35 Br	36 Kr 83.80
37 Rb 85.47	38 **Sr** 87.62	39 Y 88.91	40 Zr 91.22	41 Nb 92.91	42 Mo	43 Tc (97.91)	44 Ru 101.07	45 Rh 102.91	46 Pd 106.42	47 Ag 107.87	48 Cd	49 In 114.82	50 Sn	51 Sb 121.75	52 Te 127.60	53 I	54 Xe 131.29
55 Cs 132.91	56 Ba 137.33	57 La 138.91	72 Hf 178.49	73 Ta 180.95	74 W	75 Re 186.21	76 Os 190.23	77 Ir 192.22	78 Pt 195.08	79 Au 196.97	80 Hg 200.59	81 Tl 204.38	82 Pb 207.2	83 Bi 208.98	84 Po (208.98)	85 At (209.99)	86 Rn (222.02)
87 Fr (223.02)	88 Ra (226.03)	89 Ac (227.03)	104 Rf (261.11)	105 Ha (262.11)	106 Sg (263.12)												

Periodic Table of the Elements

58	59	60	61	62	63	64	65	66	67	68	69	70	71
Ce 140.12	Pr 140.91	Nd 144.24	Pm (144.91)	Sm 150.36	Eu 151.97	Gd 157.25	Tb 158.93	Dy 162.50	Ho 164.93	Er 167.26	Tm 168.93	Yb 173.04	Lu 174.97

90	91	92	93	94	95	96	97	98	99	100	101	102	103
Th 232.04	Pa 231.04	U 238.03	Np (237.05)	Pu (244.06)	Am (243.06)	Cm (247.07)	Bk (247.07)	Cf (251.08)	Es (252.08)	Fm (257.10)	Md (258.10)	No (259.10)	Lr (262.11)

clumps became denser and denser until eventually a very high pressure developed at the center of each clump, resulting in the combination of the nuclei of the constituent atoms and the release of large amounts of (kinetic) energy resulting in heat and radiation. Thus radiating stars were born (Some of the processes involved have been artificially induced to occur in thermonuclear weapons.). Stars formed (and continue to form) with a variety of masses. In the interior of stars which are not very massive (such as our own sun and lighter stars) the nuclear 'burning' associated with the combination of atomic nuclei only continues up to the formation iron and then stops (The fusion reactions occurring in stars are summarized in Fig. 2.2). The star then dies slowly and in a relatively uneventful way as a white dwarf. As we are interested in how the chemical elements of life form, and become available for planets, this sequence of events is not very relevant, because the heavier elements in Table 2.1 were not formed and what elements were formed remained mainly within the interior of these low mass stars.

However, for more massive stars (5 times as massive as our sun or more), something more spectacular happens: When the pressure inside the star has induced nuclear fusion up to iron nuclei and the star thus runs out of 'fuel' it implodes very fast (in a few seconds) and then a lot of material bounces off the core and explodes into space.

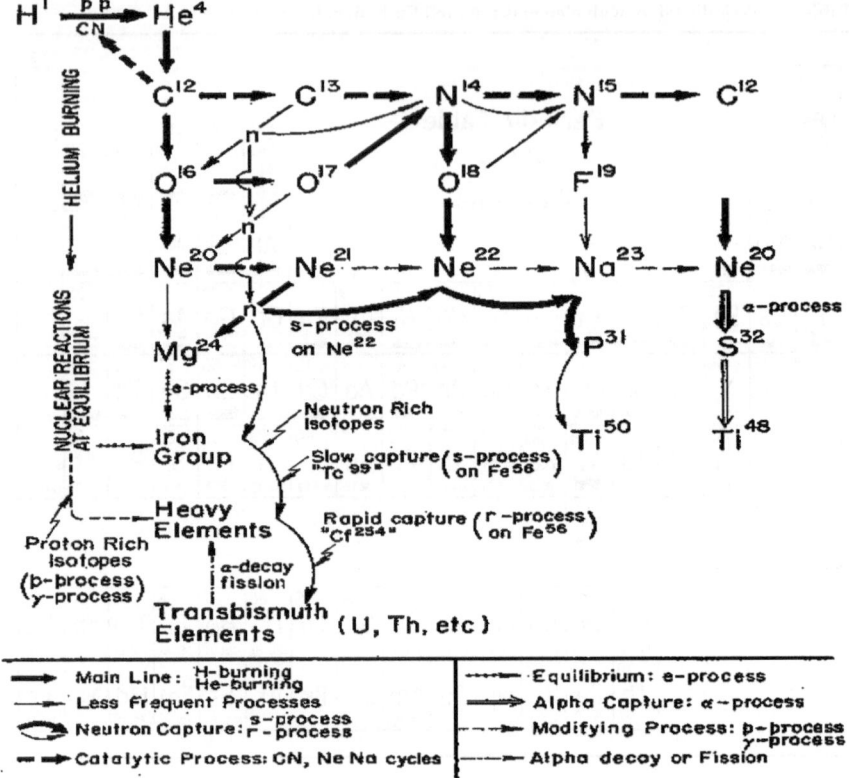

Fig. 2.2 Nuclear reactions occurring inside stars to make the elements of the periodic table, starting with hydrogen. Moving from *left* to *right* in the diagram, one usually adds neutrons to the nuclei, though in some cases, protons are added. The *superscripts* on the symbols indicate the number of protons plus the number of neutrons in the nucleus. Moving *down* the table one adds 'alpha particles' to the nuclei. These reactions are helium nuclei with 2 protons and 2 neutrons. The reactions *down* to iron all occur inside stars, but the reactions leading to the elements *above* iron are believed to require stellar explosions (*supernovae*) to occur. Aspects of the nature of the reactions leading to the higher atomic number elements above iron are still under study. From Ref. [1]

This is a 'supernova'. It results in a flash of light in the sky. The changes in color and brightness which occur during the lifetime of such a massive star are illustrated in Fig. 2.3. Supernovae big enough and close enough to earth to be observed without a telescope have been observed every few hundred years by humans and were recorded as long ago as 1054 by Chinese astronomers.

Many more supernovae are observed, in our own and other galaxies, with telescopes. (There are two types of supernovae. This is a description of type II supernovae. Type I supernovae are basically similar but occur in slightly less massive stars, between 1.4 and 5 times the mass of our sun and require accretion of mass from some outside source to implode. We will not be concerned with the distinction

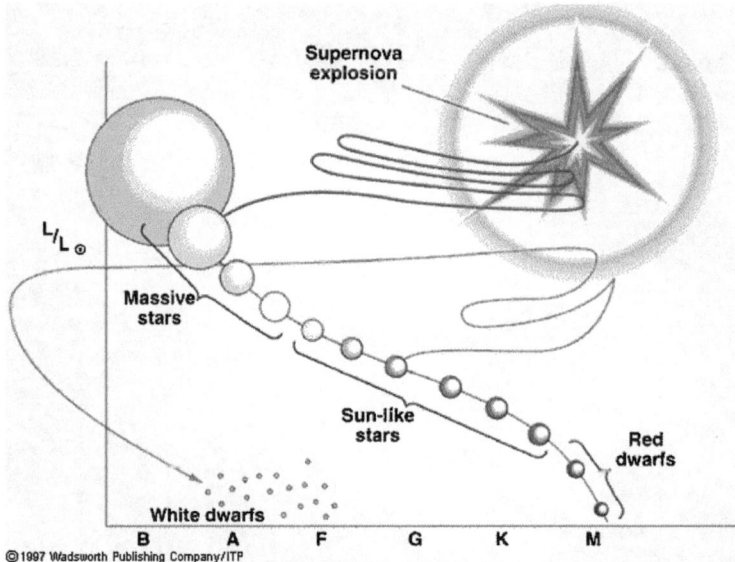

Fig. 2.3 The path which the theory of stellar evolution says that a star of mass more than 5 solar masses will follow in the so called HR diagram used by astronomers to classify stars. The *vertical axis* is the luminosity (brightness of the light) and the *horizontal axis* is the surface temperature (inferred from the color). These high mass stars live about 10^7 years. A star like our sun has luminosity 1.0 and is of type G and is on the main sequence. It is expected to live about 10^{10} years and to end as a white dwarf star, not as a supernova. From Ref. [2]

between type I and type II supernovae any more here.) The glowing remanents of a supernova which occurred in 1054 and was observed by Chinese astronomers can still be seen in the Crab nebula, of which a picture appears in Fig. 2.4.

The supernovae are significant because they produce the elements above iron in the periodic table and because they spread the products of nuclear burning in the interior of stars into the interstellar medium. Assuming that, like life on earth, other life would require that these heavier elements be present, we will expect to find life originating only on planets formed from the debris of these supernovae explosions. Such planets exist because there is evidence that, after the explosions, the material from the explosion again starts to clump up under the action of the mutual gravitational attraction of the atoms in the gas of the debris. New, 'second generation' stars are formed (and around them, at least sometimes, planetary systems which we discuss later). Such stars, of which our sun is one, reveal that they are second generation because the atoms in the gas of the stars emit light characteristic of the heavy elements which can only be formed in supernovae explosions.

With this picture and a little quantitative information about the stars we can get an estimate of the factors $N_{gal} f_{star}$ in the Drake equation. The average number N_{gal} of stars per galaxy is roughly known (this is just a matter of counting) and we will use the number 10^{11} for it. To find the rate at which matter is released by supernovae

Fig. 2.4 The Crab nebula, identified as the remanent of the supernova observed by Chinese astronomers in 1054. It lies about 6,300 light-years and is expanding at an average speed of 1,800 km/s. The image is a composite taken with the Hubble space telescope [3]

explosions in a galaxy we use the fact that supernovae occur roughly every 10^2 years per galaxy. (In our own galaxy, supernovae have been definitely observed in 1006, 1054, 1572 and 1604. However, some may have been missed and others obscured from view by the dense material at the center of the galaxy. The estimate of one every century takes into account observations of supernovae in many other galaxies as well.) Here, as in most other quantitative arguments in this book, we will only cite numbers up to the nearest power of 10. This is justified by the fact that many of the quantities we need are not known to much better accuracy than this and also by the interesting fact that estimates made using such numbers usually give better results than one might expect. Errors made in one factor of a calculation by rounding off to the nearest power of 10 tend to be compensated by errors in the other direction in other factors in the calculations. The theory of stellar evolution, amply confirmed by extensive calculations and observations, gives a lifetime for stars in the mass range leading to supernovae (more than five times the mass of the sun) which is of order 10^7 years. From these numbers we can estimate the average number N_{big} of massive stars in a galaxy which will become supernovae:

$$N_{big}/(10^7 \, yr) = 1/10^2 \, yr$$

so that $N_{big} \approx 10^5$. Since each of these stars contains at least 5 solar masses of mass, one is producing at least 5×10^{-2} solar masses of material for second generation stars per year. Once formed, these second generation stars, which are mainly of 1 solar mass each or less, live about 10^{10} years. We can estimate the number of second generation stars N_2 by equating their death rate to their birth rate:

$$N_2/10^{10}\,year = 5 \times 10^{-2}\,per\,year$$

so that $N_2 \approx 5 \times 10^8$ and the fraction f_{star} of stars in the galaxy which are second generation and of a mass similar to that of our sun is $f_{star} \approx 5 \times 10^8/10^{11} = 5 \times 10^{-3}$. This estimate is about 10 times lower than a similar one due to Hart [4]. Our considerations neglect some effects. For example, the gas from a supernova could all coagulate into another giant star which could in turn explode and so on. During the lifetime of the universe ($14 \times 10^9\,year$) there could in principle be enough time for more than 100 generations of stars if this process repeated itself. However, because only one in a million stars is big enough to make a supernova, this multiple generation process is very unlikely and most stars will be first or second generation. Also, we have restricted attention to second generation stars with lifetimes of order $10^{10}\,year$. Less massive stars than the sun will live longer and more massive ones will have shorter lives. There are reasons associated with need for a habitable climate to restrict attention to the stars with masses close to that of the sun, as we will discuss in the next part. Notice that though the lifetime of the universe ($14 \times 10^9\,year$) seems very long, it is not much longer than the estimated lifetime of our sun. On the time scales which interest us, our universe will turn out to appear to have lived quite a short time from several points of view.

In summary our estimates for the first two, astrophysical, factors in the Drake equation are

$$N_{gal} \approx 10^{11}$$

and

$$f_{star} \approx 10^{-2}.$$

References

1. E. Margaret Burbidge, G.R. Burbidge, W.A. Fowler, F. Hoyle, Synthesis of the elements in stars. Rev. Mod. Phys. **29**, 547 (1957) Copyright (1957) by the American physical society used by permission
2. Image from Michael Seeds, Horizons: Exploring the Universe, 5th Edition, Wadsworth Publishing Co. (1997), Fig. 10-1 by permission
3. Image by NASA, ESA and Allison Loll and Jeff Hester (Arizona State University) by permission
4. M.H. Hart, in *Extraterrestrials, Where are they?* 2nd edn. ed. by B. Zuckerman, M.H. Hart (Cambridge Press, Cambridge, 1982), p. 218

Chapter 3
Planetary Considerations

Now we consider the planetary factor in the Drake equation:

$$N_{civ} = N_{gal} f_{star} f_{planet} f_{life}$$

$\uparrow$

We are assuming that life will only originate on a planet somewhat like our own. We will make some more specific assumptions later on in this section, particularly with regard to the presence of water. Even without such specific assumptions, a look at our own solar system, immediately gives an idea of what can go wrong with regard to the habitability of a planet, even if a planetary system is formed around a second generation star like our own: Mercury is too hot to sustain our kind of life, Venus has a crushing pressure, a very high temperature and a sulfuric acid atmosphere, Mars is very cold with an extremely dilute atmosphere. The rest of the planets (Jupiter, Saturn, Neptune, Uranus) have no solid surface at all, are at cryogenic temperatures and feature huge gravitational fields at their surfaces. Some of moons of Jupiter and Saturn are regarded as having some possibility of harboring life, though none has been found to date and they are all at least as inhospitable as Mars. (More details concerning the properties of the planets and moons of our solar system are given in Chap. 8.) One is left with the impression that earth and its inhabitants have been very lucky. We will attempt to quantify this impression somewhat here as we estimate the factor f_{planet} in the Drake equation.

As mentioned in the introduction, it is convenient to separate the discussion of f_{planet} into two factors, which, following Hart [1], we can write as $f_{planet} = f_{PS} f_{hab}$. The factor f_{PS} is the probability that any planetary system at all forms around a second generation star and the factor f_{hab} is the probability that, given that a planetary system has formed, it contains a 'habitable' planet on which life can evolve. The scientific considerations associated with estimating these two factors are quite distinct.

J. W. Halley, *How Likely is Extraterrestrial Life?*, SpringerBriefs in Astronomy, DOI: 10.1007/978-3-642-22754-7_3, © The Author(s) 2012

Fig. 3.1 Reflected light intensity as a function of wavelength for various planets in the solar system, compared with the light intensity from the sun. Note the *log scale*. The solar light is 10^5 times brighter. From Ref. [4]

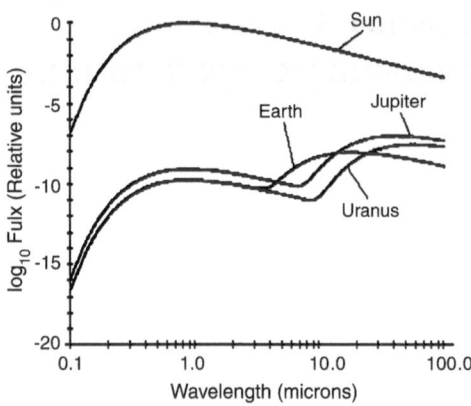

3.1 Frequency f_{PS} of Planetary Systems

We begin with f_{PS}. From the observational point of view, the science of planetary systems around other stars is very young. No planets at all had been reliably observed around any star except our sun until about 1996. This late start is associated with the fact that the electromagnetic signals from a planet orbiting around a star are extremely small compared to those coming from the star itself. This is illustrated in Fig. 3.1, which quantitatively compares the intensity of radiation from our sun and from some of the planets orbiting it as function of the wavelength of the radiation. The signal from earth is about 10^5 times smaller than that from the sun at all wavelengths and could not be detected by our optical telescopes until very recently [2, 3] (Fig. 3.2).

This problem was finally overcome in the 1990s by several indirect methods of which the most successful is the observation of the Doppler shift of the visible light from the star which results from the motion of a planet around it. The orbiting planet slightly moves the star to and fro as it goes around. As a result, the star is periodically moving away from and toward an observer so that the wavelength of the light which the star emits is shifted to the blue while the star is tugged toward the observer and toward the red as the star is pulled away from the observer. The period of the wavelength variation is the period of the orbit of the planet. By use of Newton's equations, using the period and amplitude of the wavelength shifts and assuming that there is only one orbiting planet, one quite easily obtains information about the mass of the planet times the sine of the angle i that the orbit makes with a plane perpendicular to line between the observer and the star (Appendix 3.1).

In 2011, 505 planets had been detected by this Doppler shift method. In the case of single planets, the technique permits determination of orbital radius (semimajor axis) and eccentricity and planetary mass times the sine of the angle of inclination i of the planetary orbit to the line of sight. In cases in which the orbit is almost parallel to the line of sight, so that one is observing the orbit 'edge on', one can complement the Doppler shift method by another observational technique in which the modulation of the star's light intensity is modulated as the planet passes in front of the star.

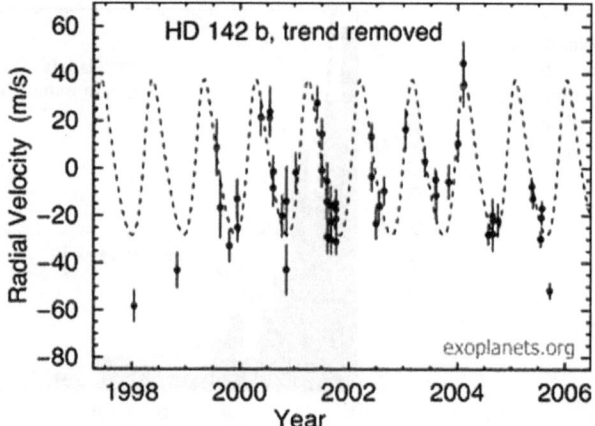

Fig. 3.2 The component of the velocity of a star that is in the direction of the earth observer as estimated from the Doppler shift. From Ref. [5] The *number* in the title is the catalogue number of the star. The *dashed* line shows the expected velocity for the orbital parameters which give the best match to the observations. See (Appendix 3.1) and [5] for more details

In those cases, termed detection of transit, it is possible to determine the mass of the planet. This had been done for about 55 planetary systems in January 2009. It has been possible to get some information about the atmospheres of the planets observed by this occultation technique as well. In some cases, complex time dependence of the wavelength shifts indicates the presence of more than one planet, and more compli- cated calculations have been used to deduce the nature of the planetary systems. The discovery of planets outside our solar system is occurring very rapidly at the time of writing. For a review of the field up to 2005 see Ref. [6]. For a current encyclopedia of the planets (505 in April 2011) discovered by various means see Ref. [7] and also [8]. Of particular interest is the Kepler satellite, described in [9] which is espe- cially designed to find earth-like planets using the transit technique. It had reported [5] confirmed planets in February 2011 (and 1,235 'candidates'). The mass distribution of discovered planets as of 2008 is shown in Fig. 3.3.

We note the following: Most of the masses of planets discovered to date are much larger than those of the earth (which is about 0.003 Jupiter masses). This is because the current detection methods have only been able to detect a few planets with masses near the mass of the earth. In many cases, however, the existence of earthlike planets in the same planetary systems in which heavy planets with masses comparable to and exceeding the mass of Jupiter is not excluded by the data. Most of the detected planets are likely to be gaseous ones like our Jupiter and not particularly hospitable to life. Theoretical models of planetary formation [11] suggest that formation of large gas like planets by gravitational coagulation of dust disks around stars is a less likely process than the formation of smaller earth like planets. Therefore the current searches could be missing earthlike planets around many of the stars which have been studied. The mass distribution of the observed extrasolar planets encourages

Fig. 3.3 Mass distribution of observed extrasolar planets (in 2008 from Ref. [10]. Figure used by permission). At the time, 207 planets within 200 pc of the earth were included

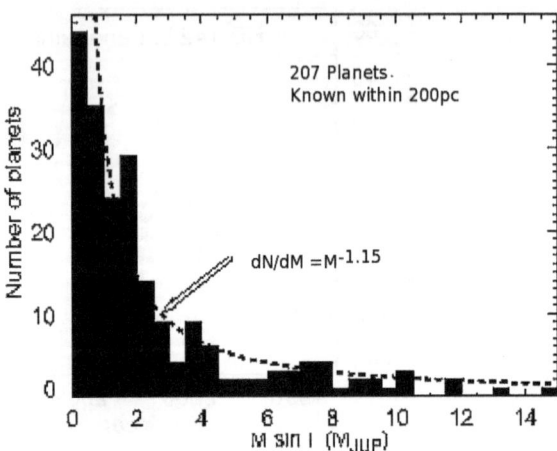

this idea (Fig. 3.3) because there is a sharp rise toward the low mass side of the distribution.

On the other hand, the existence of Jupiter like planets may be essential to the existence of smaller, habitable planets like the earth, because numerical simulations [12] suggest that Jupiter protects the earth from frequent meteoric bombardment.

The fraction of stars f_{PS} which have planets can still be only roughly estimated from the available information. For example, in 2002, one could get an estimate f_{PS} from the available information using the fact that 1,000 stars had yielded 93 planetary systems. These 1,000 stars were reported to be essentially all the sunlike stars within 30 pc of earth (This is consistent with our other estimates supposing that the galaxy has a volume of around 10^{12} pc^3). Thus, if this part of the galaxy is typical, we would find $f_{PS} \approx 10^{-1}$. A more careful estimate [13] made at about the same time resulted in the conclusion that "at least 9% of Sun-like stars have planets in the mass and orbital period ranges M sin i > 0 : 3M$_{Jup}$ and P < 13 years and at least 22% have planets in the larger range M sin i > 0 : 1M$_{Jup}$ and P < 60 years". These numbers are roughly consistent with the first estimate.

Estimating f_{PS} from available observations of planetary systems is probably more reliable than using estimates based on theories of planetary formation [11]. However, from theoretical models one can get some insight into how this number could be small: Most stars seem to start life with dust disks swirling around them, from which planets are believed to form. However, simulations of the process by which the dust in these disks coagulates into planets suggest that if the density of dust is too high, the result is the formation of a second star, instead of a planetary system. This second star revolves around the first star in a binary system. Indeed, more than half of the stars in the sky are observed to be binaries. Binaries tend to efficiently sweep up the remaining dust into the stars themselves and planetary systems are unlikely around them. In fact, in our own solar system, Jupiter, is not very far from the mass which would have made it a second star in our system.

On the other hand if the density of dust in the disk is too low, the simulations show that giant planets of the sort detected in the current searches cannot form. Thus there is a narrow window of dust densities in which planetary systems containing giant planets can form. This effect could result in a number as small as 10% of stars having giant planets orbiting around them.

3.2 Fraction f_{hab} of Planets Which are Habitable

The next problem is to estimate the probability that these planetary systems contain habitable planets. How we define habitable will depend on our assumptions about the requirements for (and definition of) life. Assuming as before that the life we are considering is chemically similar to our own we will require temperatures and atmospheric pressures in a quite narrow range in which water is liquid. A phase diagram, showing the values of temperature and pressure for which liquid water exists, appears in the next figure [14]. (There are about 10^5 Pa in one atmosphere of pressure, while one atmosphere is close to the average atmospheric pressure of air on the surface of the earth. The temperature scale is in absolute degrees Kelvin. One obtains Celsius degrees by subtracting 273 from these numbers.) The mean surface temperature of the earth is 288 K and its pressure is around 10^5 Pa. Thus we lie in the lower left hand corner of the pressure temperature region in which water is liquid. Note that the pressure scale is logarithmic, so the allowed pressure range is considerably larger than the allowed temperature range (Fig. 3.4).

It is not known whether the restriction to the liquid water range is too restrictive or not. NASA currently is sponsoring a program to explore the extremes of temperature, humidity and pressure in which bacterial forms of terrestrial life can exist. It has been established that bacteria can be kept outside the liquid water range for extended periods and then returned to it and revived. One form of life in extreme environments is the fauna living in and near thermal plumes on the ocean floor. There are thriving communities [16] at depths of 8,000 ft (approximately 2×10^6 Pa.) The plumes have temperatures of up to 670 K degrees, though most of the organisms are reported to be living at much lower temperatures. This pressure is in the liquid water range. The highest temperature is to the right of the critical point in the phase diagram, in the region of 'supercritical' water where gas cannot be distinguished from liquid. Many strange forms of life are observed near these vents in the ocean trenches. An example is shown in Fig. 3.5.

One may also question whether water is required as a solvent for life. To my knowledge, no form of terrestrial life exists without any water at all. Whether similar biochemistry could occur without water as a solvent is unknown, but none has been produced artificially in a laboratory or found on another planet. Further, one can understand why water is a particularly, possibly uniquely, appropriate solvent for life. It is an almost universal solvent (meaning that it dissolves nearly everything), a property closely associated with the large electrical dipole moment of its constituent molecules and its high dielectric constant. Water is neither acidic nor basic

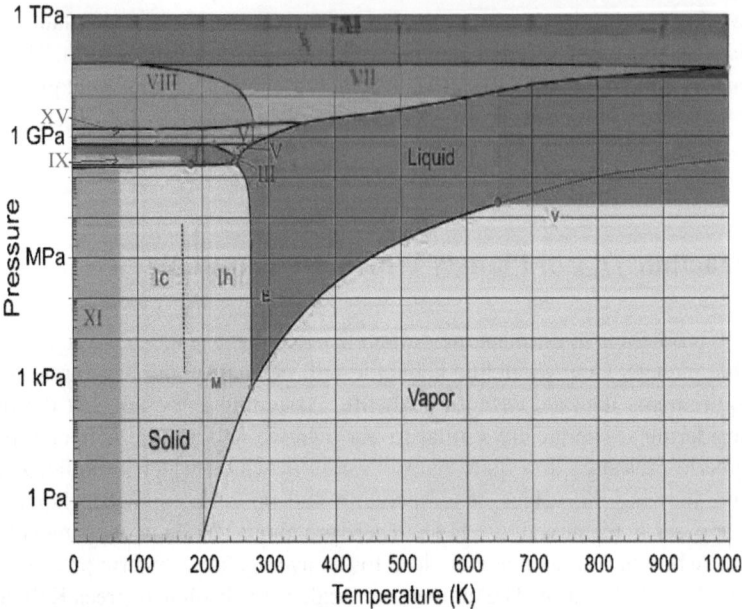

Fig. 3.4 Phase diagram of water. The E is at the approximate temperature and pressure of the earth's surface. V and M label the corresponding points for Venus and Mars. From Ref. [15]

Fig. 3.5 A 'tube worm colony' near an ocean trench. The tube worms are up to 1.5 m long. The image was taken by the deep-submergence vehicle Alvin at 2,510 m depth of a deep-sea hydrothermal vent community. Specifically this an assemblage of the tubeworm community, Riftia pachyptila, was taken in December, 1993, following an April 1991/1992 eruption that created the venting habitat (one of the first documented deep-sea eruptions on a mid-ocean ridge), Ref. [17]

(pH = 7 or neutral) and it has a high surface tension, enhancing capillary action essential for transport of fluid through capillaries in organisms. Water has a very high specific heat, making it very useful as a thermal buffer for maintaining a constant temperature in organisms (and in human produced engines of various kinds). No other liquid with this combination of properties is known by chemists. While these considerations do not prove that a life-like chemistry could not exist without water, they do suggest it. In view of our limited knowledge of this entire aspect of the question, we will assume that the existence of liquid water is required for the kind of life like entities we seek. However other solvents have been considered as possible media for life-like systems. Among those often mentioned are liquid ammonia (NH_3), hydrogen cyanide (HCN), hydroflouric acid (HF), hydrogen sulfide (H_2S), methanol (CH_3OH) and hydrazine (N_2H_4). Many of their properties and a discussion of their suitability as media for life-like chemistry are reviewed in Ref. [18].

We have almost no knowledge of the atmospheric conditions on the extrasolar planets (a few observations have very recently been reported which have some bearing on this) so we will be required to depend mainly on theoretical considerations and on data from earth and other solar planets. Unfortunately, the theoretical modeling of climate, which is extremely important for efforts to determine appropriate public policies concerning anthroprogenic gas emissions, for example, is notoriously difficult. Attempts to model the climate of the earth over millions of years have been made. A great deal is known about the earth's climatic history [19], for example from analysis of the chemical content of ice cores taken at great depth near the earth's poles, and this information can be used to assess the reliability of the modeling methods. During such long periods, the chemical content, as well as the pressure and temperature, of the oceans and atmosphere of the earth are known to have changed very significantly and climate models are required to reproduce these changes which occurred in the past on earth. However even when this requirement is met, the predictive power of existing computer models of climate is uncertain, as evidenced by the fact that they do not all predict the same future for earth's climate. (The International Panel on Climate Change dealt with this problem by averaging results from the available climate models in making its predictions and recommendations.) We will not be able to go into much detail concerning climate modeling here, and it is not evident that doing so would significantly improve our estimates of the probability of existence of habitable planets. This is because models are having difficulty producing reliable and reproducible estimates of global climate change on scales of a few to hundreds of thousands of years whereas we will need to estimate something about the climate over times at least of the order of the hundreds of millions of years which elapsed during the latest stages of evolution on earth.

To get an estimate without going into all the complexities of climate modeling, we discuss some of the basic physics which determines planetary temperatures here and use them to get some useful bounds on habitability based on the criterion of the existence of liquid water. It can be argued that the simple considerations we use here will overestimate the probability of habitability, because more detailed climate models tend to be extremely sensitive to initial conditions: little changes in the way a planet starts its life can make an enormous difference in whether it turns out to be habitable or not. (This feature of theoretical models is sometimes called 'chaotic' but this word should not be overinterpreted in this context.) Since most conceivable conditions will not be habitable, more complicated models which take account of this 'chaotic' feature, as our argument below does not, are likely to give lower probabilities of habitability than we will find.

Climate models begin with the determination of a 'black body' temperature for the planet, which is based on a simplified model in which the planet simply absorbs radiant electromagnetic energy from its star and reradiates electromagnetic energy as if it were a 'black body' (without reflectivity) and characterized by an absolute surface temperature T. Such a black body is known to radiate energy at a rate σT^4 per unit surface area where σ is a known constant determinable from the fundamental constants of nature and equal to 5.68×10^{-8} J/s m^2 K^4. Suppose we have a star of luminosity (radiated energy output per unit time) L which is a distance R from a planet of radius r. The energy absorbed from the star by this energy output is

$$energy\ in/time = (\pi r^2/4\pi R^2)L$$

whereas the energy radiated by the planet at temperature T is

$$energy\ out/time = 4\pi r^2 \sigma T^4$$

We are assuming that the temperature over the surface of the planet is uniform, which of course it is not. This is a simplified model. After some time the surface temperature of the planet will adjust itself so that the energy input is exactly balanced by the energy output giving

$$(\pi r^2/4\pi R^2)L = 4\pi r^2 \sigma T^4$$

which can be solved for the temperature T:

$$T = (L/16\pi R^2 \sigma)^{1/4}$$

This quantity is called the black body temperature of the planet. The quantity $L/4\pi R^2$ is the amount of energy per unit time and per unit area which a detector at a distance R from the star would record. For the earth, the quantity $L/4\pi R^2$ is called the solar constant and it has been quite carefully measured to be 1,360 J/s m^2 giving a surface temperature for the earth from the model of

$$T_{earth,model} = (1,360\ \text{J/s}/4 \times 5.68 \times 10^{-8}\ \text{J/s m}^2\ \text{K}^4)^{1/4} = 278\ \text{K}$$

The average surface temperature of the earth is in fact 288 K so this is remarkably close to the right answer. However one gets a hint that this is a dangerously simple model by carrying out the same exercise for Venus. The model temperature is

$$T_{venus,model} = (150/108)^{1/2} \times 278 = 327\ \text{K}$$

(The ratio 150/108 is the ratio of the radius of the earth's orbit to the radius of Venus' orbit.) However the measured surface temperature of Venus is over 740 K!. The bad result for Venus is mainly due to the 'greenhouse effect': The visible light from the sun goes through the atmosphere of Venus more easily than the redder and invisible infrared radiated from the planet gets out through the atmosphere. As a result Venus has a higher surface temperature than our simple model would indicate. It is exactly this greenhouse effect which may be warming our earth due to human additions of infrared absorbing gases like CO_2 to the atmosphere and which many fear could lead to catastrophic warming of the earth.

For Mars the same model gives

$$T_{Mars,model} = (150/228)^{1/2} x278 = 225\,K$$

The real, measured mean surface temperature of Mars is reported to vary between about 190 and 270 K so the model works a little better here. However it has been suggested that on Mars an opposite, runaway glaciation could be taking place in which the planet gets colder than predicted due to the precipitation of snow (made of carbon dioxide on Mars) which reflects the sunlight more than a black body so the planet cools, resulting in more precipitation of snow and further cooling etc.

Despite the problems, we will use the black body model to try to estimate f_{hab} because the black body model works moderately well and because vastly more elaborate models are both too complicated to use and not terribly reliable. We will suppose that we are considering stars with masses near that of the sun. As mentioned earlier, most stars are in this mass range, though the heavier ones leading to supernovae play an important role in producing needed elements they are not very numerous (one in a million stars). Furthermore those massive stars have very short lives (10^7 years or so) which would not leave enough time for evolution to take place on their planets even in the most optimistic evolutionary scenarios. Now we ask the question, over what range of distances of a planet from a star of mass similar to our own sun would liquid water be possible?

Looking at the phase diagram of water, one sees that the requirement that liquid water be present puts quite a definite lower bound on the temperature of around 270 K so that the water does not freeze. But the upper bound on the temperature, determined by the requirement that water does not all vaporize, depends on the pressure of the atmosphere at the surface. We do not know any simple way to predict the surface pressure in general. For example, the pressure at the surface of Venus is 90 atmospheres but all the professional scientists failed so badly to predict it before it was measured that the first planetary probe on the surface of Venus was crushed by the atmosphere. Constraints on the pressure don't seem to be provided by the habitability requirement either: life has been observed into the temperature and pressure region associated with supercritical water above the critical point. To make some progress, we will tentatively suppose that the upper limit on the temperature which would allow life on the surface of a planet is similar to the temperature of the thermal plumes near the ridges at the bottom of the oceans on earth, where life is found under the most extreme conditions known. This assumption gives about 700 K as the upper limit on the temperature. Now we can work backward through the black body model to get the range of distances R from a star over which a planet might be habitable by some form of life. Notice that the distances in the model don't depend on the radius of the planet.

$$270\,\mathrm{K} < 278/R^{1/2} < 700$$

or

$$0.157\,\mathrm{AU} < R < 1.06\,\mathrm{AU}$$

where the distances are expressed in 'astronomical units' in which the distance of the earth from the sun is 1 AU. Because we decided that life might tolerate a very hot planet like Venus (based on the thermal plume data) we are predicting that life might exist on a planet very close to a sun like star. Even for Mercury, $R = 58/150 = 0.39$ AU so this is a very liberal criterion. This estimate assumes we are dealing with a sunlike star. It can be refined by taking the observed distribution of stellar luminosities into account, but such refinements are not justified given the uncertainties in the other assumptions, particularly regarding the upper limit on the temperature.

Now to estimate f_{hab} we need to decide how likely it is to find earth like planets in this range of distances. The existing data on observed planets is not of direct use because almost all the observed planets are much too high in mass and will be huge gaseous Jupiter like objects. We can, however, probably exclude the systems which have Jupiter like objects inside the needed range because the Jupiter like objects might be expected to prevent formation of earth like objects in the same range. It is not at all clear how far out we should require Jupiter like objects to be. Let us suppose that they must be at least 2 AU away from the star, to leave room for an earthlike planet in an inner orbit. (Jupiter is at 5.2 AU). Then, in 2002, 68 of the 93 planetary systems discovered would be excluded from the set which would allow a habitable earthlike planet to form. If at least one earth like planet always forms (this seems likely if a Jupiter like planet forms) and if it were equally likely to form an earth like planet anywhere inside 2 AU and if a Jupiter like planet is required (to protect the earth like planet from meteors as discussed earlier) then the probability is very roughly estimated as $\approx 1 - 68/93 \approx 0.3$. (A correction to exclude earth like planets very close to the star has a negligible effect on the estimate.) Recall that our criterion includes Venus and Mercury as habitable. We have taken no explicit account of pressure or of required chemistry. Without detailed modeling it seems prudent to assume that the possibility that these factors do not come out favorably would reduce the likelihood of habitability by as much as another factor of ten. We end up with $f_{hab} \approx 10^{-1}$ but this is much less certain than our previous estimates so we will write $f_{hab} \approx 10^{-1\pm1}$.

References

1. M.H. Hart, in *Extraterrestrials, Where are they?* ed. by B. Zuckerman, M.H. Hart, 2nd edn. (Cambridge Press, Cambridge, 1995), p. 218
2. P. Kalas, J.R. Graham, E. Chiang, M.P. Fitzgerald, M. Clampin, E.S. Kite, K. Stapelfeldt, C. Marois, J. Krish, Science **322**, 1345–1348 (2008)
3. C. Marois, B. Macintosh, T. Barman, B. Zuckerman, I. Song, J. Patience, D. Lafrenière, R. Doyon, Science **322**, 1348–1352 (2008)
4. B.F. Burke (eds.), *Towards Other Planetary Systems* (NASA Solar System Exploration Division, Washington D.C., 1992) available at http://www.stsci.edu/ftp/ExInEd/electronic-reports-folder/TOPS.pdf
5. J.T. Wright, O. Fakhouri, G.W. Marcy, E. Han, Y. Feng, J.A. Johnson, A.W. Howard, D.A. Fischer, J.A. Valenti, J. Anderson, N. Piskunov, lanl.arXiv.org/astro-ph/arXiv:1012.5676 (2010) using the Exoplanet Orbit Database and the Exoplanet Data Explorer at exoplanets.org
6. B. Reipurth, D. Jewitt, K. Keil (eds.), *Protostars and Planets V* (University of Arizona Space Science Series, Tuscon, 2007)
7. http://exoplanet.eu/catalog.php
8. http://exoplanet.eu/searches.php
9. http://exoplanet.eu/searches.php
10. G.W. Marcy, R.P. Butler, S.S. Vogt, D.A. Fischer, J.T. Wright, J.A. Johnson, C.G. Tinney, H.R.A. Jones, B.D. Carter, J. Bailey, S.J. O'Toole, S. Upadhyay, Physica Scripta. T130, 014001 (2008)
11. J.L. Lunine, in *Extraterrestrials, Where are they?* ed. by B. Zuckerman, M.H. Hart, 2nd edn. (Cambridge Press, Cambridge, 1995), p.192
12. G.W. Wetherill, Lunar Plant. Sci. Conf. **24**, 1511–1512 (1993)
13. C.H. Lineweaver, D. Grether, Astrophys. J. **598**, 1350–1360 (2003)
14. http://www.sbu.ac.uk/water/chaplin.html
15. M. Chaplin,Water Phase Diagram. http://www.lsbu.ac.uk/water/phase.html by permission
16. http://pubs.usgs.gov/gip/dynamic/exploring.html#anchor14337915
17. Tim Shank (Woods Hole Oceanographic Institution), by permission
18. D. Schulze-Makuch, L.N. Irwin, *Life in the Universe* (Springer, Berlin, 2004) Chapter 6
19. N. Christie-Blick, Icarus **50**, 423–443 (1982)

Chapter 4
Biological Factors

We come next to the factor f_{life} in the Drake equation.

$$N_{civ} = N_{gal} f_{star} f_{planet} f_{life}$$
$$\uparrow$$

This is the most difficult factor to estimate, mainly because the detailed mechanism of the origin of life on earth is not known. A somewhat less obvious problem is that estimates of the factor f_{life} are very sensitive to the criteria which are used to identify the presence of life. Attempts to identify the essential features required to label a material system as 'alive' are numerous, and necessarily arbitrary because the word 'life' can be defined by its user in many ways. In the first part of this chapter we will discuss the problem of estimating f_{life} assuming that its existence requires the presence of some form of information carrying polymer which retains the information needed for the living system to function, as DNA does in terrestrial life. In the random polymer assembly model to be described, we further assume that this information carrying entity is assembled first during life's origin. We will see that these assumptions, if correct, lead to very small values for f_{life}. Next we discuss various ways which have been proposed for removing the constraints imposed by these assumptions. With sufficient broadening of the definition of life, we will see that much larger values of f_{life} can be derived, however at the price of allowing entities into the category of living systems which many people would not regard as 'alive'.

If we wish to use the Drake equation to estimate the probability of observing civilizations then we can factor f_{life}, which is then the probability that a 'civilization' exists on a habitable planet, into two factors: the first factor, which we term f_{prebio} is the probability that the biochemical elements necessary for the beginning of biological selection and evolution appear. If we are only using the Drake equation to estimate the number of life systems, then this is the only factor needed. The second factor, termed f_{evol} is the probability that biological evolution, once started, leads to a complex system which will meet our (still vague) criterion of 'civilization'. We are so far assuming here that if life appears on a planet, it is there because it has evolved

J. W. Halley, *How Likely is Extraterrestrial Life?*, SpringerBriefs in Astronomy,
DOI: 10.1007/978-3-642-22754-7_4, © The Author(s) 2012

there. This is not necessary. It is entirely possible to imagine that life might migrate from one planet to another either on the same star or (much less likely) to other stars and that this could occur either spontaneously or through the intentional activities of the life on the source planet. We will discuss the possibility of migration in some detail in Chap. 6.

4.1 The Probability f_{prebio} That Life Begins on a Habitable Planet

To discuss f_{prebio} we will briefly review what is known about the biochemical nature of life on earth. As noted earlier, it is possible to imagine that the biochemistry of life elsewhere might be different. However there are two reasons for thinking that this possibility might not affect our estimates as much as it might initially appear. Firstly, detailed studies of alternative chemistries, extending over nearly a century, have not yielded any alternative forms of life, and have revealed in increasing detail how specifically favorable the existing carbon based chemistry is for life. (This was first discussed by Henderson [1]).

However, it is possible that life evolved as it has on earth to take advantage of the presence of abundant carbon and that in other planetary environments, evolution would use other chemistries. Further, the arguments for carbon and water as essential may be presuming a definition of what we mean by 'life' which is unnecessarily parochial. Nevertheless, we will see that the feature of biochemistry which dominates our estimate of f_{prebio} is the requisite complexity or the closely related information content of the needed biomolecules. If we suppose that a similar complexity will be required for life to evolve from alternative biochemistries, then including them in our estimates would only involve multiplying by the number of viable alternative chemistries. If the number of possible alternative chemistries is not large, then the estimate of the probability f_{prebio} will not be hugely affected by taking them into account. Unfortunately, estimating the number of viable alternative chemistries when we have not yet found any is obviously going to be difficult and it is not obvious that it is small. We will take up the issue of the number of possible alternative chemistries later in the chapter.

All known life is based on a genetic code realized in one or more polymeric (chain) molecules constructed from five nucleotides (of which only four are used in any particular chain molecule). The atomic structure of these nucleotides is shown in Fig. 4.1. These components of the genetic code molecules are termed cytosine (C), uracil (U) or thymine (T), adenine (A) and guanine (G). The resulting long chain molecules are termed ribonucleic acids (RNA) or deoxyribonucleic acids (DNA). The code is carried from one generation of organisms to the next by DNA, which is a double helical molecule (Fig. 4.2) in which G is always paired to C and A is paired to T.

This pairing permits the DNA double helix to reproduce itself by separation and the replating of a new partner helix from the surrounding soup containing nucleotides in

Fig. 4.1 Nucleotide building blocks of the DNA molcule. The distances between the atoms in this and the following two figures is of the order of a few times 10^{-10} m. Figure from reference [2]

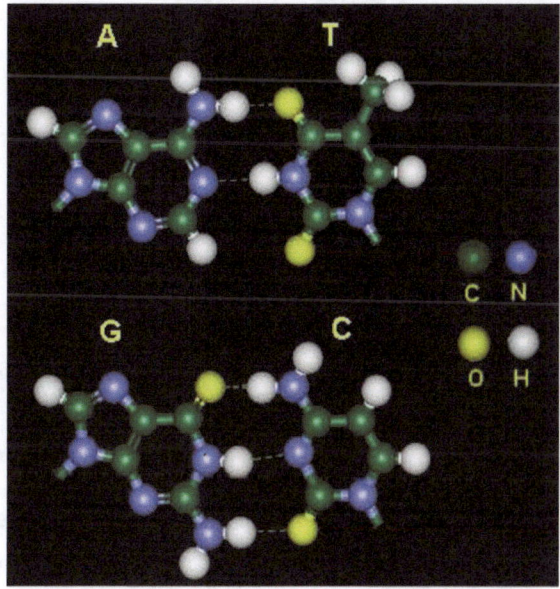

obedience to the pairing rules. A key function of the DNA is to direct the formation of proteins, which carry out most of the mechanical functions of the organism. Proteins are also long polymeric molecules. For proteins, the building blocks are amino acids, of which 20 are used in life on earth as illustrated in Figs. 4.3, 4.4. The DNA codes protein structure by means of a genetic code in which 3 nucleotides are used to indicate one amino acid as described in Fig. 4.5.

One of the vital functions of proteins is to act as enzymes, which are biochemical catalysts holding relevant constituents of biochemical reactions in place so that the reaction can occur rapidly and in the appropriate way. In particular, the processes involved in the decoding of DNA to form proteins is guided by a huge (on atomic scales) enzymatic 'factory' called a ribosome which is made of proteins and RNA. (An X-ray derived picture of a ribosome is shown in the Fig. 4.6).

For our purposes, it is quite important to understand the atomic size of these constituents of life as well as the accuracy with which the reproductive process needs to take place. (Other organic cell constituents, such as cell walls made of fatty lipids, seem to form quite easily and spontaneously. They are not likely to constitute the 'bottleneck' to getting evolution started.) In humans, the DNA has recently been mapped (approximately) in the Human Genome Project and found to contain 3.2×10^9 base pairs (the number of nucleotides along one strand of the double helix). For a single celled bacterium such as E Coli the corresponding number is about 5×10^6. There is evidence that such single celled bacteria were present on earth more than 3 billion years ago. The simplest known organisms are viruses consisting of a DNA molecule and a protein coat. However viruses, such as the T2 virus which played an important role in the history of molecular biology,

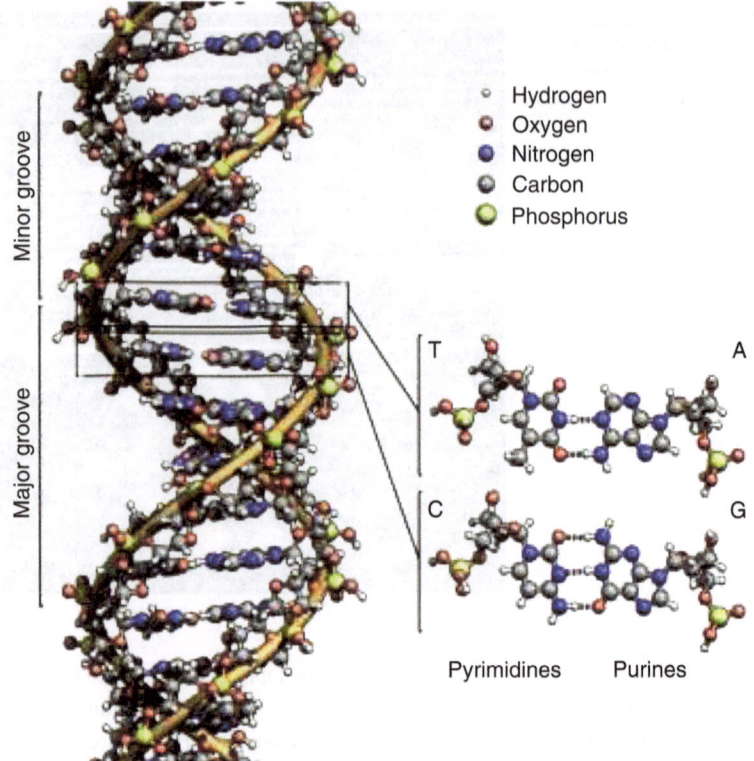

Hydrogen
Oxygen
Nitrogen
Carbon
Phosphorus

T   A

C   G

Pyrimidines   Purines

Fig. 4.2 Portion of a DNA double helix. The image is from [3]

Fig. 4.3 The structure of amino acids, the fundamental building blocks of proteins [4]. The group labelled R may be any one of the twenty shown in Fig. 4.4

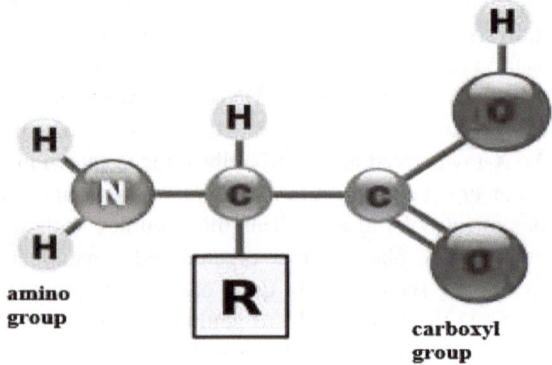

amino group

R

carboxyl group

General structure of an
α - amino acid

are parasitic on single celled bacteria in that they could not reproduce unless they had the bacterial ribosomes and other features of the bacterial cell to assist them. It

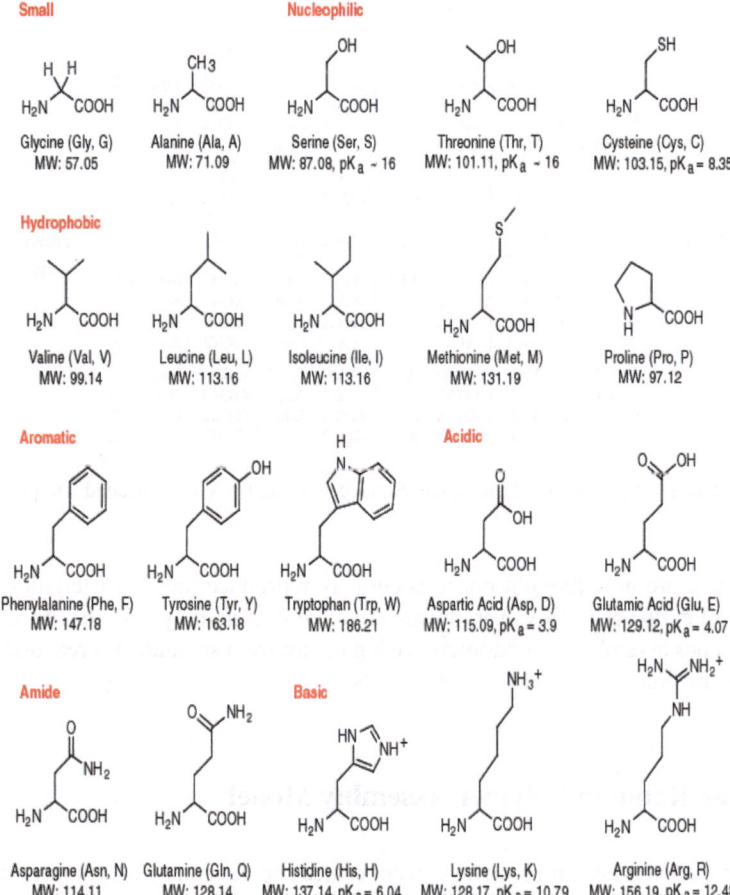

Fig. 4.4 The twenty R groups of the amino acids which appear in biological proteins. Reprinted from http://www.neb.com (2011) with permission from New England Biolabs

is found that transcription of the genetic code does not need to be perfect for life to propagate. It is estimated that only about 10% of the DNA base pairs must be exactly fixed in the code in order to get functioning proteins for life processes.

A common qualitative idea about how life began on earth is that the right mix of atomic constituents (produced by stellar processes as we discussed earlier) was present in the oceans, possibly in shallow ponds, and that over time, random chemical reactions resulted in the biochemistry required for reproductive processes to begin. Some experiments have been carried out to test aspects of this idea. One of the first experiments was done by Stanley Miller [5]. More recent work is described in Ref. [6]. By subjecting a mixture of CH_4, NH_3, H_2 and H_2O to electric discharges simulating lightning, for example, it was possible to produce amino acids in a rather short time. Adenine has also been formed in this way. The chemical conditions of the Miller

Second letter

		U	C	A	G	
	U	UUU Phe UUC Phe UUA Leu UUG Leu	UCU Ser UCC Ser UCA Ser UCG Ser	UAU Tyr UAC Tyr UAA Stop UAG Stop	UGU Cys UGC Cys UGA Stop UGG Trp	U C G A
	C	CUU Leu CUC Leu CUA Leu CUG Leu	CCU Pro CCC Pro CCA Pro CCG Pro	CAU His CAC His CAA Gln CAG Gln	CGU Arg CGC Arg CGA Arg CGG Arg	U C G A
First Letter	A	AUU Ile AUC Ile AUA Ile AUG Met	ACU Thr ACC Thr ACA Thr ACG Thr	AAU Asn AAC Asn AAA Lys AAG Lys	AGU Ser AGC Ser AGA Arg AGG Arg	U C G A
	G	GUU Val GUC Val GUA Val GUG Val	GCU Ala GCC Ala GCA Ala GCG Ala	GAU Asp GAC Asp GAA Glu GAG Glu	GGU Gly GGC Gly GGA Gly GGG Gly	U C G A

Third Letter

Fig. 4.5 The genetic code by which triplets of nucleotides on RNA are translated into proteins

experiment are now thought not to accurately reproduce the characteristics of the surface of early earth. However, the idea that the starting polymers for life arose by spontaneous assembly of monomers arising in turn from spontaneous reactions in an aqueous medium on the early earth persists.

4.2 The Random Polymer Assembly Model

If the nucleotides and amino acids can be formed spontaneously in this way, then the next step is to randomly assemble long chain polymers with sequences of them long and accurate enough to permit some sort of reproductive life cycle to begin. Though in existing life much more is required (see Fig. 4.6) than the presence of the DNA, it is argued that perhaps RNA alone, which is capable of playing some of the roles that proteins do in modern life, might alone form a reproducing life like system and thus the formation of the initial RNA might suffice to start the process. To illustrate some features of the difficulty involved in this idea, suppose that we want to produce the messenger RNA of a single celled organism like E Coli. The evidence indicates that such organisms, called prokaryotic, appeared on earth more than 3.5 billion years ago and were the only types of organisms present until about 2 billion years ago, when protozoa, still single celled but with nuclei, appeared. Multicelled organisms seem to have appeared quite suddenly (on geological time scales) a little more than a billion years after that [9].

DNA (and mRNA) of the bacterium E Coli has 5×10^6 base pairs, but let's say only 1/10 of them need to be exactly right so we need to assemble a molecule with exactly the right nucleotide at 5×10^5 positions. We imagine doing this by picking nucleotides out of the primeval ocean at random and adding them to a chain

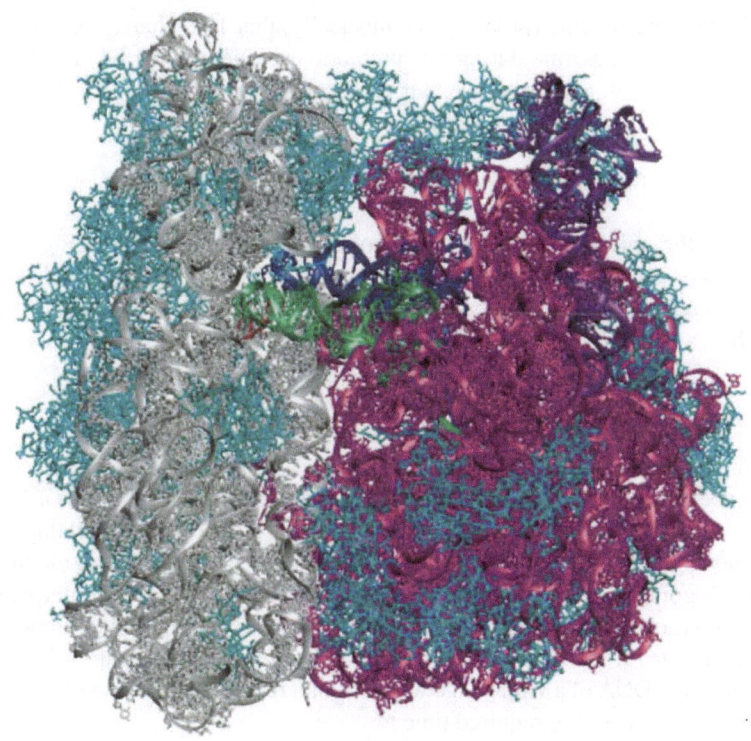

Fig. 4.6 X-ray derived picture of the structure of the ribosome of the organism Thermus Thermophilus [7]. The ribosome consists of both RNA and proteins and is required in all existing organisms for translation of the DNA code from messenger RNA into proteins. This ribosome contains more than 100,000 atoms. Figure from reference [8]

we are building. (The random processes of fluctuating motions in the ocean would presumably do it this way.) Considering that there are 4 nucleotides (we ignore the complication that thymine and uracil play similar roles) we have one chance in 4 of getting the first base right. We have 1 chance in 4^2 of getting the second base right. (We assume it doesn't matter what goes between the first and second base that are fixed. Actually, how many pairs go between the first and the second one might matter even if their identity does not, but this consideration will not affect the result much.) Similarly we have $1/4^3$ chance of getting the third base right and $1/4^{500,000}$ chance of getting the whole thing right. The last number is immensely small. (Your calculator will overflow if you try to calculate it.) Let us forge ahead to an actual estimate of probability. This tiny chance of getting the right polymer means that, in order to succeed by a random process, on average $(1/2) \times 4^{500,000}$ polymers will have to be constructed. How long would this take? One can imagine that this is not happening with just one polymer but with a very large number of them in the ocean. Estimates of the number of carbon atoms in the primeval oceans are of the order of

10^{44}. Suppose we assume (probably optimistically) that 10^{42} chains are being built simultaneously all the time during this prebiotic evolutionary period. The reaction time associated with adding a base pair can be assumed to be of the order of 1 ms (again optimistically). Thus our estimate of the time required is

$$(1/2)1 \times 500,000 \times 4^{500,000} \times 10^{-3}/10^{42}\,\text{s} \approx 10^{300,960}\,\text{s} \approx 10^{300953}\,years$$

Recall that the universe itself has an age of about 10^{10} years. Assuming that the planet we are considering has lived a substantial portion of that time (which is not too unreasonable) then if the probability of success in the prebiotic phase of evolution would be

$$f_{prebio} \approx 10^{10}/10^{300953} = 10^{-300943}$$

assuming that life began by this random polymeric assembly process. This extremely tiny number, if correct, means that the Drake equation is going to give us a number for N_{civ} which is very much less than one. In fact the number is so small that even a calculation of the number of civilizations in the entire observable universe (about 10^{12} galaxies) is going to give a number much much less than 1. This can be interpreted to mean that our own existence is extremely improbable. (Not impossible. Improbable events do occur.) If the needed 'starter genome' were 1,000 base pairs long (much shorter than the DNA of a viral phage which cannot reproduce on its own) this would reduce the estimate of the required time to

$$(1/2)1000 \times 4^{1000} \times 10^{-3}/10^{42}\,\text{s} = 10^{562}\,\text{s} \approx 10^{555}\,years$$

Though this is a huge change in the numbers, the qualitative conclusion is very similar.

The result in the last paragraph is so astonishing that the first reaction of most scientists to it is that there must be some mistake. (A much more complete discussion of this problem appears in the first section of Ref. [10]. The problem has sometimes been called the 'evolutionary paradox'.) Indeed much of the effort in scientific studies aimed at understanding the origins of life are directed at finding alternatives to the naive random polymer assembly model we have described.

The most obvious way to make the numbers more reasonable is to suppose that we have overestimated the length of the polymer required to get the reproductive biology and associated Darwinian selection started. For example, it has been speculated in some detail by Francis Crick and coworkers [11] that a reproductive cycle with chemistry related to that of living organisms might be sustained with only RNA molecules. RNA molecules are an attractive choice because they can act both as genomes, carrying information, and as enzymes (biological catalysts) so some kind of life-like autocatalytic chemical cycles might, in principle proceed without the involvement of proteins.

The paper by Crick and coworkers does not report an estimate of the minimum required length of the tRNA needed to start the process. However some experiments with RNA by the group of Eigen [12, 13] have shown that some features of autocatalytic reproduction can be made to occur in the laboratory using rather short RNA's of the order of 10^2 base pairs in length. Unfortunately an enzyme (a catalytic protein of biological origin) of a much larger size was needed to assist the process, so this is not a complete answer to the question. Another problem with the Eigen experiments is that, though the RNA molecules started to evolve toward larger sizes in the experiments, they were observed to reach a maximum size much less than typical sizes in organisms in our biosphere and then to stop evolving. This was interpreted as due to the 'mutational load' on the longer RNA, due, for example to damage by cosmic rays. This load increases with polymer length, and, without the error correction mechanisms present in living organisms, it causes the longer RNA molecules in the Eigen experiments to die out. Experiments with even shorter RNA's of order 10 nucleotides long have also been reported, but these usually also require support from large enzymes.

It not is clear that such an 'RNA world' would leave any paleontological signature and none has been found. Neither have such autocatalytic cycles involving short RNAs been found in any natural systems on earth in modern times. It is sometimes suggested that they do not appear because such early stages of life would be 'eaten' or competitively overwhelmed by the more complex organisms which succeeded them. This seems possible, although the present biosphere sustains a wide variety of organisms including many primitive ones, so to demonstrate such extinction requires a more detailed description than is presently available.

Another popular idea for resolving the 'evolutionary paradox' is that catalysis greatly speeds up the reaction rate so that combinations can be tried more rapidly. Typically (Appendix 4.1) catalysis, which has been proposed by Cairns-Smith in the present case to occur at mineral surfaces, speeds up reactions by factors of the order of 10^6. Reduction by six orders of magnitude in the time required to assemble a starter genome would not change the qualitative conclusion leading to the evolutionary paradox. Cairns-Smith [14] proposed that the first reproducing entities on earth were crystals (which do spontaneously assemble from solution quite quickly) and that these somehow catalysed the formation of self replicating biopolymers. (See also Ref. [15].) There might possibly be some hint of this in crustaceans, in which the opposite effect takes place (catalysis of mineral shells by biopolymers) and crustaceans do appear early in the fossil record. Whether such mechanisms could decrease the estimated time for prebiotic evolution by several thousand orders of magnitude is unknown. It seems superficially unlikely but it is certainly worth exploring.

It is possible that reproducing entities can start with even smaller molecules. We can turn the problem around and ask how small the polymer must be to reduce the required time to that apparently observed on earth: Let the number of fixed base pairs N. Then we would require

$$t_{average} = N \times 4^N \times 10^{-3}/10^{42}\, s \approx 10^9 \, years \qquad (4.1)$$

giving $N \approx 100$ (Fig. 4.7). Perhaps this is not impossible and it is the size of the RNA molecules in the Eigen experiments mentioned above (which were noted to

Fig. 4.7 Polymer assembly times estimated in the random polymer assembly model, as a function of the length of the required starting polymer. The dashed lines indicate the age of the earth and the observable universe

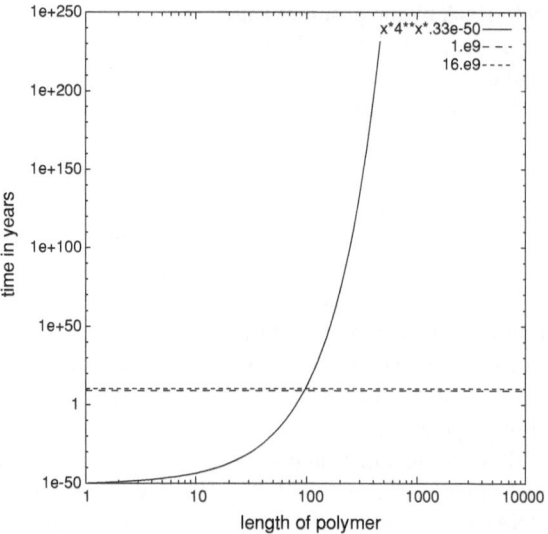

also require the presence of a much larger biological enzyme). No such prebiotic mechanisms have been detected in the natural world on earth.

In conclusion with regard to f_{prebio}, if we assume that the critical step in getting life started is the random assembly of a 'starter genome' we are finding that the value of f_{prebio} depends very strongly on how long that genome needs to be. If it must be like those in the unicellular organisms on earth today, then f_{prebio} would have to be of order $10^{-301000}$. If they are of the order of 1,000 fixed base pairs then f_{prebio} is of order 10^{-500}. If they are of order 100 fixed base pairs then $f_{prebio} \approx 1$. The absence of biochemical evidence in the natural world for reproduction and other lifelike features of molecules at the shorter end of the scale inclines us to regard the smaller values of f_{prebio} as somewhat more probable.

Let us provisionally suppose, within the random polymer assembly model, that our own existence, *in the observable universe* is near the value to be expected statistically (rather than being a very improbable event). This provisional assumption that our existence is not improbable is closely akin to a controversial idea called the 'anthropic principle' which is not part of normal scientific discussion of these issues and which will be further discussed in Chap. 9. Using this assumption, and by the same arguments and models as before, using our previously determined values for the other factors in the Drake equation, taking $f_{evol} \approx 1$ (see below) and assuming 10^{12} galaxies in the observable universe we find

$$10^{23}(5x10^{-3})(10^{-1})(10^{-2})f_{prebio}(1) \geq 1$$

so that $f_{prebio} \geq .2x10^{-17}$ Putting this back in the random polymer assembly model we get the needed biopolymer length N from

$$N \times 4^N \leq 10^{10}\, yr x (3 \times 10^7\, s/yr)/(10^{-3}/10^{42}\, s)x(2 \times 10^{-18}) \approx 10^{80}$$

or $N \approx 130$ corresponding to a time for polymer assembly of about 10^{28} *year* on average. We are pushed toward the conclusion, that, despite our lack of a detailed biochemical model, the minimum fixed base length of the starting biopolymers was not more than a few hundred if life is not a rare event. The conclusion is illustrated in another way by plotting the estimated time for random production of an initiating biopolymer of length N as a function of N as shown in Fig. 4.7.

Quantitatively, these arguments leave the following ranges for the length N of the starting polymer and for f_{prebio}, assuming the random polymer assembly model and that catalytic enhancements of biopolymer assembly rates cannot increase them by 100s of orders of magnitude:

1. If $N \leq 100$ then $f_{prebio} > 10^{-6}$ and we expect more than one lifesystem per galaxy.
2. If $N \approx 100$ then $f_{prebio} \approx 10^{-6}$ (giving one life system per galaxy)
3. If $N \approx 120$ then $f_{prebio} \approx 10^{-17}$ (giving one lifesystem per universe)
4. If $N > 150$ then the expected number of life systems in the observable universe is much less than 1 and our existence is an improbable event.

Alternative 1 may lead to problems with the Fermi paradox as discussed in Part II of the book. Alternatives 1–3 leave the problem that no biochemical systems with any of the properties of life (reproduction, metabolism, etc.) are known with information codes this short.

Within the 'random polymer assembly model' for prebiotic evolution, the following possibilities, then remain open, assuming that still undiscovered catalytic mechanisms do not enhance the biopolymer assembly rate by several hundred orders of magnitude:

1. Life appeared on earth at approximately the average expected time Then if there are one or more civilizations per galaxy, the starting biopolymer is roughly 100 monomers or less in length and if there are one or more civilizations in the observable universe then the starting biopolymer is 120 monomers or less in length. No currently observed organisms have DNA this short and for independent organisms the shortest DNA is of order 10^4 monomers in length.
2. Life appeared on earth at a time much less than the average expected time. This possibility is statistically consistent since we have only one experimental instance in the data set.

4.3 Other Models for Prebiotic Evolution

There are several alternative scenarios to the random polymer assembly model for the prebiotic stage of evolution. Generically, these alternative approaches are sometimes called 'metabolism first' or 'proteins first' models, whereas the random polymer assembly model is the simplest of a class of models in which the essential initiating event is the assembly of a gene, hence it is one of a category of models called 'genetics first' or 'genome first' for the origin of life. However a clearer distinction between the random polymer assembly model and the models to be discussed here is that the alternative models do not assume that there is just one, or a very small number, of successful chemical ways to get life started. In the random polymer assembly model as we described it, there was just one way out of 4^N possibilities, to get the right starter molecule. This is not unreasonable if one is starting with RNA, because biologically, organisms are very intolerant of errors in their genes and have evolved elaborate mechanisms to avoid errors during reproduction. The supporters of the 'metabolism/proteins first' alternatives contend that these error intolerant genes evolved late in the evolution of life and that, prior to that, a much more 'sloppy' form of reproduction evolved. In that early stage, cycles of autocatalyzed protein reactions, or metabolic cycles of reactions like those involving ATP in contemporary cells (or both) are thought to have appeared, without the guidance of a genome. Although there is paleontological evidence of bacteria-like life on earth from 3.8 billion years ago, it is not actually possible to determine whether the fossil cells contained genes or not, although it is often assumed that they did. So the proposed scenario is not eliminated by the existing paleontological data.

One of the earliest such models is called 'The Hypercycle' by its originator, Eigen [16] who describes a rather detailed scenario, involving cycles of proteins followed by introduction of RNA and finally DNA. As noted above, attempts in Eigen's laboratory to reproduce features of this scenario have been partly successful.

Another, more abstract model is due to Freeman Dyson [17]. In Dyson's model, a simple starting 'soup' is postulated, in which molecules form at random by addition of monomers, much as in the random polymer assembly model. As in almost every theory of life's origin, an enclosing cell is assumed to exist, inside which these assembly processes occur. The spontaneous assembly of cell like micelles is well known in chemistry and is fast, so this is generally agreed not to be a central issue [18]. A key difference between Dyson's model and the random polymer assembly model is that he essentially assumes that the number of 'successful' ways to assembly the molecule is very large. Consider the time to assemble a successful starter molecule (or collection of molecules-Dyson does not distinguish whether it's one or a collection). Suppose that there are N_{ways} different ways to do the assembly and, as before n^N possible forms of the molecule altogether. Here n is the number of types of monomer which are being used in the assembly. For nucleic acids which we assumed were in use in the description of the random polymer assembly model, $n = 4$ whereas for proteins $n = 20$. As before, N is the number of molecules in the assembled entity. Now the time for assembly, per cell, will be

$$time / cell = \tau n^N / N_{ways}$$

and for all the cells simultaneously assembling

$$time = \tau n^N / N_{ways} / (number\ of\ cells)$$

This should be compared with (4.1). τ is the time per addition of a molecule to the assembly, which we took in (4.1) to be 1 ms. We took $n = 4$ and the number of cells to be 10^{42}. Apart from the unimportant factor $N/2$ in (4.1) the two relations are the same when $N_{ways} = 1$. An important innovation introduced by Dyson is to suppose that, unlike modern life, in early life one only needed to get N_a of the molecules right in order to get successful reproduction, where $N_a \approx 0.8N$. But most importantly he assumed that the assembled molecule would 'work' no matter what the order of the correct N_a monomers. Now the number of ways to choose N_a objects out of N, if you don't care about their order is

$$N!/(N_a)!(N - N_a)! \approx (x^{-x}(1 - x)^{x-1})^N$$

where $x = N_a/N$ is the fraction of the monomers which you have to get right. If $x = 1/2$ then this factor is 2^N so the time required would increase by $(n/2)^N$ instead of as n^N due to this factor. (By further requiring that N not be too large (of order 10^4) and including some effects of autocatalysis in the model, Dyson concludes that the times required with $x = 0.8$ are short enough so that life by such a model could get started without a rare event.

Another version of a 'metabolism/proteins first' model was proposed by Stuart Kauffman [19, 20] Kauffmann also does not assume that a DNA molecule with a particular code is required for the initiation of life. Instead Kauffman considers a system of many polymers, one of which catalyses a reaction producing a second polymer. The second polymer in turn catalyses a reaction producing a third polymer and so forth. If this chain eventually closes so that the Mth polymer catalyses the production of the first, then one has a closed autocatalytic cycle of reactions. Such closed, autocatalytic cycles are known in experimental chemistry [21], though most of them do not involve polymers. In his simplified computer models, Kauffman shows that when the numbers of different types of polymers gets large enough, then the formation of such autocatalytic cycles becomes very probable. Taking the view that the formation of such autocatalytic cycles is the essential biochemical feature of life, Kauffman suggests that this scenario may contain the essential elements of the process which led to the initiation of life and may resolve the evolutionary paradox.

In comparing existing life to the Kauffman scenario, if one focusses attention on the nucleic acids, there does not seem to be a very clear similarity. DNA reproduces itself directly and not through a cycle of intermediate steps involving other polymers. One the other hand, as we have emphasized, this reproduction of DNA is assisted by a very complex system of enzymes in cells, and one may suggest that this system of enzymatic proteins plus DNA may be acting somewhat like a Kauffman auto-catalytic system, and this is quite close to what Eigen has proposed. Alternatively, as suggested by Dyson, autocatalytic cycles involving proteins might have evolved first, followed later by the development of genes within the already reproducing cells. In the terrestrial biosphere, the study of protein control networks which regulate the activation of genes is currently being elucidated in great detail [22]. In the scenarios suggested by Kauffman, Dyson, Eigen and others it is hypothesized that the appearance of the control network occurred *prior to* the appearance of the gene because the appearance of such a network is statistically much more probable than the appearance of a functioning 'naked' gene. The gene is hypothesized to have evolved later, within the network, as a convenient bookkeeping device.

A 'metabolism first' approach is represented by Wachterhauser [23] who proposes that metabolic cycles processing energy in the cells evolved before the DNA-RNA-protein cycle appeared. In fact metabolic chemical cycles in existing organisms seem much like what is to be expected in a Kauffman-type of model of origin. There is a large number of chemical participants forming a cycle with no particular molecular participant which can be characterized as carrying information which guides the process. The fact (Fig. 4.8) that more than one metabolic cycle appears in existing organisms suggests that initiation of these cycles is more probable than initiation of the genomic system . It seems quite plausible that the existing metabolic cycles may have evolved through a mechanism like that studied by Kauffman and that, as his models would suggest, such evolution is relatively probable.

In these and several other theoretical scenarios directed at the solution of the problem posed by the random polymer assembly model, one deals with the central problem by assuming that the system can get a metastable steady state started in many ways. Instead of needing to assemble just one genome out of 4^N possible ones, one allows many initial states. Such scenarios could resolve the difficulty. From the point of view of the search for extraterrestrial life, we may ask to what extent such a diversity of initial states will converge through evolution on the same or very similar biospheres at later times. This question can be posed in terms of what are called 'basins of attraction' in complex dynamical systems [24]. In many mathematical models of such systems one finds that, for a range of initial conditions the system converges on the same final state (which may be a dynamical 'limit cycle') as time evolves. But outside that set of initial conditions it may converge to a completely different state or may not converge to any steady state at all.

Then the issue which is posed by origin of life models with a range of possible initial conditions is: 'Do all these allowed initial conditions lead to life-like systems with the same fundametal chemistry, or are they sensitive to initial conditions so that the resulting biochemistries of evolved life-like systems are very diverse?' If the answer to that question is that, in the real systems, the diverse initial conditions which could start a life-like metastable system result in many different biochemistries, then it could turn out that terrestrial biochemistry provides very few hints concerning what should be sought in a chemical search for extraterrestrial life. In Appendix 4.2 I describe some results from a simplified Kauffman model which suggest that the resulting biochemistries might indeed be very diverse.

At a more concrete level, there are not many, if any, experimental laboratory systems which reproduce the conjectured spontaneous autocatalytic cycles and it is obviously important to find them if this class of ideas is to be regarded as useful for understanding the origin of life. If they are found, then one can consider whether they lead to one biochemistry or to many. If they were to all lead to the same or very similar biochemistries, then there is a distinct danger of explaining too much, in the sense that life with biochemistry very similar to ours might be predicted to be very probable on earthlike planets. Such a conclusion might be inconsistent with observation, since setting $f_{life} = 1$ leads to a million biospheres in the galaxy and existing observational data may rule out such a large number as discussed in part II of this book. On the other hand if the spontaneously generated initial 'sloppy' biochemistries could produce

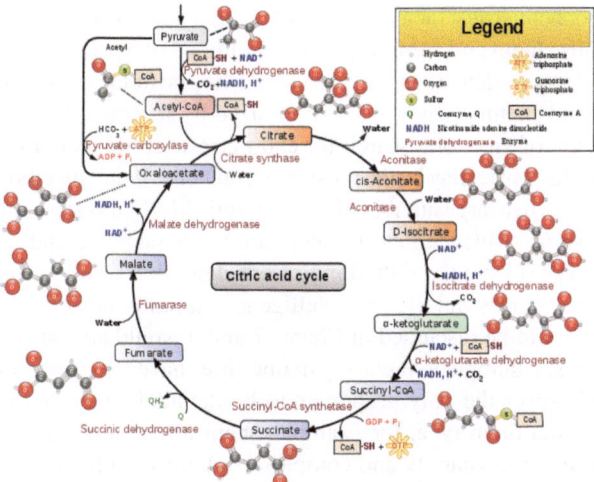

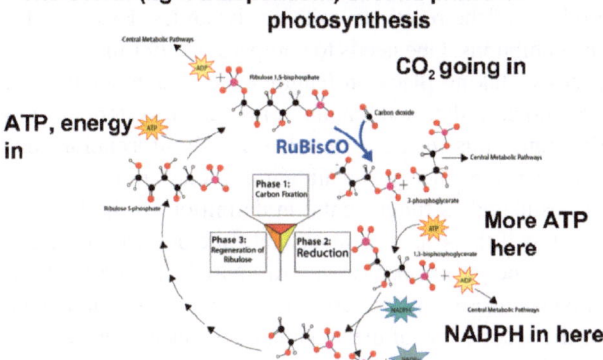

Fig. 4.8 Two metabolic cycles observed in terrestrial organisms. The citric acid, or Krebs cycle (*top*) occurs in mitchondria. It starts with pyruvate (*top of diagram*) and produces CO_2 and water, while adding energy to the energy carrying molecules NADH, FADH and ATP. It is observed to run backward in some anaerobic bacteria [25]. The Calvin cycle (*bottom*) occurs in plants. It is the second stage of photosynthesis, and occurs in the absence of light. The Calvin cycle consumes CO_2 from the atmosphere and uses the energetic molecules ATP and NADH produced by light during the first stage of photosynthesis to produce energetic glucose. The sugars produced by this cycle are broken down and used in the Krebs cycle in mitochondria of the cells of animals which consume the plants. (RuBisCo is an enzyme which catalyses the first step in the Calvin cycle. This is a simplified picture and details, particularly in Phase 3, are lacking). Figure by Michael D Jones, MSc by permission

very diverse biochemistries, then it is more likely that human observation might have failed to recognise them and there is reduced likelihood of a paradox. The possibility that existing efforts to detect extraterrestrial life has postulated too limited a range of complexity and chemistry has been emphasized by several authors [26, 27].

From the point of view of a search for extraterrestrial life and/or extraterrestrial intelligence, the last issue may be the most important one. If the 'metabolism/proteins first' scenarios lead to a large number of autocatalytic life-like systems in the universe (though we have no evidence that it does) then to characterize and discover them will require a very substantial rethinking of our definition of life. The approach to the search for extraterrestrial life and intelligence currently taken by NASA probes and SETI searches to be discussed in Chaps. 7 and 8 would almost certainly fail to find such systems. Attempts to precisely define 'life' have a long and indeterminant history [28]. However the only properties to be definitely expected from life-like systems of the Kauffman type, for example, seem to be autocatalysis, (implicitly dynamic) chemical metastablity and complexity. A metastable system of chemical reactions is a dynamical system of reactions which continues for a long time, utilizing free energy from some outside source, without decaying to contain only the lowest free energy products of the reactions, with no dynamics. Exactly what is meant by 'a long time' is ambiguous. One needs to compare the lifetime with relaxation times of internal processes taking place on length scales much smaller than the size of the system as a whole. In thermodynamic terms such a system has low entropy and is not in equilibrium. Thus such a system has a high information content and must utilize some source of free energy to maintain its metastability. These last properties are sometimes postulated as fundamental in definitions of lifelike systems but one sees that they follow if the system is metastable and autocatalytic. The requirement of a source of free energy has been much discussed. It is certainly necessary, but, because the universe as a whole is very far from thermodynamic equilibrium (this is the origin of the second law of thermodynamics) such sources of free energy are quite common and the availability of free energy is unlikely to be the limiting factor in development of lifelike systems. What might be meant by complexity is actually a very subtle question and we will return to it, particularly in Chap. 7.

Several authors have used the word 'order' to describe the life-like final state in models of prebiotic evolution such as Kauffman's. But this word does not seem to me to convey much further information about the meaning of the phenomon envisioned. In fact, in condensed matter physics, 'order' usually refers to essentially static phenomena such as crystalline or magnetic ordering of atomic characteristics in space. Such states are metastable, but these ordered states of condensed matter physics are not macroscopically dynamic or autocatalytic and would not satisfy most proposed criteria of complexity. Cairns-Smith [14] has envisioned a 'living' prebiotic system on a crystal surface, but it has characteristics unlike ordinary crystals. The mathematics of Dyson's prebiotic model is similar to that used to describe a phase transition to a ferromagnet, but the 'ordered' states in the ferromagnet and envisioned in his model of prebiotic life have the differences described above.

The concept of life used, implicitly or explicitly, in models by the 'metabolism/ proteins first' schools does not appear to require individual organisms, species or reproduction. This is explicit in Dyson's discussion, where he explains that such phenomena would have to come at a later stage than the one his model is attempting to describe. In our own biosphere, it is not clear that the concepts of individual organisms and species are completely well defined. The issue of definition of species has troubled biologists at least as far back as Darwin. It remains problematic for prokaryotes and the word 'strain' is often preferred [29]. Examples such as ant colonies call the concept of individual organism into question: Should the concept be applied to the individual ants or to the colony as a whole? [30]. At the level of microbiology, it is very probable that organelles in eukaryotes (cells with nuclei) are formerly independent organisms incorporated into the eukaryotic cell during evolution by the process of endosymbiosis [31]. From this point of view, our biosphere as a whole might be regarded as living, as in the idea propagated as the 'Gaia hypothesis' [32–34] since it is certainly astoundingly metastable (3.5 billion years !), autocatalytic and complex. (The lively debate about the Gaia concept appears to be in part about the definition of life and in part about the mechanisms by which the very long metastability of the earth's biosphere has been maintained.) A similar idea appears in the work of Feinberg and Shapiro [27] who advocate searching for biospheres, not individual living species, and who describe species as an evolutionary adaptation. One may say that including 'autocatalytic' in the set (metastable, autocatalytic, complex) in this revised list of properties of life-like systems, one has called 'reproduction' by another name. In fact autocatalytic systems are reproducing themselves to maintain metastability. However this is really a generalization of the most common notion of reproduction, since it does not require any concept of species or individual organisms. Thinking in this direction, one might think more carefully about whether a phenomenon such as the 'great red spot' on Jupiter (see Fig. 4.9) might be described as 'alive' in this sense. Perhaps one would conclude that it is not sufficiently complex.

One may argue that a definition of life-like systems requiring only that they be dynamically metastable, autocatalytic and complex is insufficiently restrictive by testing it against such cases as fire, working refrigerators etc. The usefulness of such exercises is limited. However, though there are certainly borderline cases which meet the criteria but are not intuitively lifelike, one can usually see that the definition, if a little counter intuitive, is not absurd. Consider a working refrigerator. Taken alone, it is operating metastably but not autocatalytically. However if one considers a system including the power grid, the refrigerator manufacturing facilities, repair and replacement systems for the refrigerator then it seems somewhat more reasonable to regard such a system, which includes components traditionally regarded as alive, as lifelike. As another example, one can consider stars themselves. They are certainly metastable and complex, but there may be some question about whether they are autocatalytic in any well defined sense, and it is unconventional to regard them as alive. While not insisting too much on it, one can point out that the dynamics of the sun is not by any means completely random and is characterized among other regularities by an 11 year cycle and that it is maintaining its dynamical low entropy metastable state by use of its internal free energy source. The sun's expected lifetime of 10 billion years is much longer than any of the timescales characterizing processes occurring within it on length scales shorter than its size. Psychologically, most individuals will

Fig. 4.9 The Great *Red* Spot on Jupiter as photographed by the Hubble space telescope [35] in 2008. It is an anti-cyclonic storm with a diameter several times the earth's diameter which has been observed in its present state for more than 340 years. In 2005 it was observed to 'give birth' to the similar smaller anticyclone which is observable just *below* and to the *left* of the *large spot* in the picture, and then, in 2008, a *third spot* appeared which is the smaller one directly to the *left* of the *large spot*. The *intermediate size spot* has a diameter similar to that of the earth

reject the idea that the sun is alive, but it may be of some interest that this is a modern view. Ancient cultures, including very sophisticated ones, regarded the idea that the sun is alive as very natural. If one looks at the system of all stars in a galaxy, then the cycles of death and birth of stars (these are the terms that astronomers use) as discussed in Chap. 1 have some of the characteristics of an autocatalytic system.

In summary, it should be evident, as we already mentioned, that the evaluation of f_{prebio} is very sensitive to how we define life. The narrower the definition, the smaller the estimates will be. If we insist that only systems very similar to the one on earth be regarded as biospheres, then it is likely that the estimates of f_{prebio} will be very small, though how small depends on unknown details concerning the origin of life on earth. If the definition of lifelike systems is sufficiently broadened however, then f_{prebio} can be made as large as we like up to the case of the last example discussed in which the broadened definition included stars, of which there is a huge number (10^{22}) in the observable universe, and we would regard every one of them as harboring at least one 'biosphere'. This however, would take us outside the framework of the Drake equation which assumed that the life sought involves systems which exist on the surfaces of planets.

4.4 Other Arguments Relevant to Estimating f_{prebio}

Returning to more restricted definitions of 'life' there are several other arguments for a very small value of f_{prebio} one of which I will now briefly review [36, 37]: Let the time which was actually required for life to evolve to its present state on earth be

t_{earth}. From paleontological evidence, this is between 3 and 4 billion years, but we will say 10^9 years. But to the same accuracy the lifetime t_{star} a sunlike star is about 10^{10} years. Now consider the average time which would be required for complex life to evolve on an earth like planet around a sun like star (assuming that the star doesn't die, for the moment). Let it be $t_{average}$. We have been trying to estimate $t_{average}$ in our discussions of the Drake equation. But if the appearance of life on earth were an extremely improbable event, then t_{earth} could be much shorter than $t_{average}$. Now the lifetime of a sunlike star is known to depend on various properties of nuclear reactions, while $t_{average}$ depends on various features of biochemistry as we have been discussing, so it seems extremely unlikely that $t_{average}$ would turn out to be similar in magnitude to the lifetime of the star. Either $t_{average} >> t_{star}$ or $t_{average} << t_{star}$ is much more likely. But because t_{earth} is almost of the same order of magnitude as t_{star} this strongly suggests that t_{earth} is *not* of the same order as $t_{average}$. Furthermore if $t_{average} << t_{star}$ then we would be observing many civilizations in the galaxy (as we are not) and it would be very probable that life had developed more rapidly on earth than it did. We are left with the possibility that $t_{average} >> t_{star}$ and therefore that $t_{earth} << t_{average}$ so that the appearance of life on earth was a very improbable event. This is consistent with the estimates of $t_{average}$ which we got from the random polymer assembly model. However these considerations do not give a quantitative estimate for f_{prebio} (which in this language would be estimated as $10^9 year/t_{average}$.)

Though it has long been recognised by several imminent scientists [38, 36], an elementary fallacy has persisted on this subject, resulting in the assumption that $t_{earth} \approx t_{average}$ and giving very high values for the number of civilizations in the galaxy. Fundamentally this fallacy arises as a consequence of taking one instance of a distribution of events dependent on random processes and trying to draw statistical inferences from it. Here the one event is the occurence of life on earth, while the many events are multiple occurences of life on many planets. To get $t_{average}$ from observations we would need to observe life appearing on many planets and take the average over all the times required in each instance. Since we have not observed life evolving on many planets, we cannot do that. (One possible origin of the persistence of this fallacy may be the fact that in Ref. [38], where it is clearly discussed, the description of the problem is couched in the language of the so-called 'anthropic principle' which will be discussed somewhat further later in this chapter and in Chap. 9. While it is technically correct to describe the fallacy in that way, the idea of the 'anthropic principle' has been abused and misused by later writers, possibly casting suspicion on all arguments which purport to use it. There may also be some psychological reasons for the persistence of the fallacy. These are briefly discussed in Chap. 9.)

There is one other piece of information that can be used to get a little more information about the statistical possibilities: There are several lines of evidence suggesting that the initiating events in prebiotic evolution occurred only once on earth. These include the fact that all living organisms share the same genetic code, including, remarkably the code realized on DNA molecules of the same 'handedness' or chirality. It is sometimes suggested that life with alternative chemical codes or chirality might have arisen and been driven to extinction by competition with the ancestors of existing surviving organisms. While this may be possible, the argument must take account of the fact that (1) alternative chemistries are not observed to arise spontaneously over time, so the rate, if it exists, must be quite low and (2) organisms based on alternative chemistries are not likely to compete for the same types of nutrients as the organisms of the ancestors of existing life, and certainly those ancestors could not use the alternative chemistries themselves as nutrients. Thus one might expect niches for the life based on alternative chemistries, if it occurred, to survive.

If one accepts that the essential prebiotic events occurred only once on earth, then one can add the following statistical argument to the discussion: Suppose, as is the case for random uncorrelated events, that the occurence of these prebiotic events follows a Poisson distribution $P(k, t, \tau)$, which states that the probability of k events occurring within a time t is

$$P(k, t, \tau) = (1/k!)(t/\tau)^k exp(-t/\tau)$$

where τ is the average time for the event to occur. With the assumption of this paragraph, the essential prebiotic event occurred once within about 0.5 billion years and, during the subsequent 4 billion years, it did not occur again. The probability of the first event (one occurence in 0.5by) is

$$P_1 = (0.5/\tau)exp(-0.5/\tau)$$

(measuring τ in billions of years) and the probability of the second event (no prebiotic event in 4 billion years) is

$$P_2 = exp(-4.0/\tau)$$

so the probability of the two events occurring together is

$$P_1 P_2 = (0.5/\tau)exp(-4.5/\tau)$$

τ is what we would like to know. We can plot the joint probability as a function of τ as shown (Fig. 4.10).

The probability is always <0.04 which is the value it takes if τ is near the age of the earth. However, there is a very long tail giving a probability of around 1/2 a percent if τ is 100 billion years or so. If τ were really of the huge magnitudes suggested by the naive random polymer assembly model for initial polymer lengths of 1,000 units or more then this argument would also lead to the conclusion that the occurrence of life on earth must be an extremely rare event.

Fig. 4.10 Probability that the prebiotic event occured once and only once on earth, as a function of the average time for it to occur

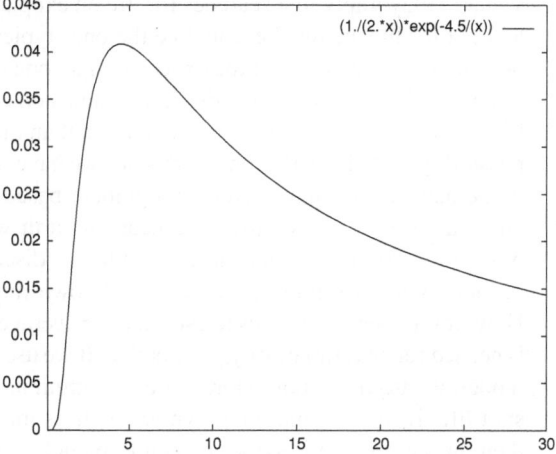

4.5 Summary of Conclusions Concerning f_{prebio}

In summary the following possibilities remain open, assuming that still undiscovered catalytic mechanisms do not enhance the biopolymer assembly rate by several hundred orders of magnitude:

1. Life appeared on earth at approximately the average expected time. Within the random polymer assembly model (also called 'genome first' or 'naked gene'), if there are one or more biospheres per galaxy, then the starting biopolymer is roughly 100 monomers or less in length and if there are one or more biospheres in the observable universe then the starting biopolymer 120 monomers or less in length. No currently organisms have DNA this short and for independent organisms the shortest DNA is more than 10^4 monomers in length. If the starting polymer were as short as 10 nucleotides, then problems would arise with regard to the 'Fermi paradox' because the random assembly model would predict very common life-like systems which are not observed. Even on earth, such a scenario would predict life-like systems with many distinct biochemistries which are not observed if the starting polymer is as small as 10 nucleotides.

2. Life was initiated on earth in the expected average time in a 'metabolism/proteins first' type of scenario. In such models it is likely that $f_{prebio} = 1$. (Dyson shows explicitly how this could arise in his version of such a model). In such a scenario, the fact ('Fermi paradox') that we have not observed life-like systems elsewhere might be explained by the hypothesis that each evolved 'biosphere' is expected to be very different chemically and in other ways from our own biosphere so that it would be extremely difficult to recognise. Within the context of such a scenario, the Fermi paradox might also be resolved if evolution (the second factor in f_{life} discussed below) usually stalled at the microbiological level leading to biospheres consisting only of microorganisms hidden beneath planetary surfaces [39]. This

scenario may imply that searches for life on exoplanets should not be confined to looking for biochemical events like the ones typical of life on earth. Depending on detailed calculations of the expected time, one could run into trouble with the failure to observe alternative life-like systems on earth itself in this scenario.

3. Life appeared on earth at a time much less than the average expected time. This possibility is statistically consistent since we have only one experimental instance in the data set. We have given an argument based on comparing the lifetime of our star to the time for life to appear on earth which appears to support this possibility. This possibility can account for the discrepancy between the estimates of biopolymer assembly times and the known time for life to appear on earth. However it does not help us to estimate the average time for life to appear, which is needed for an estimate of f_{life}. For that, if we use the random polymer assembly model, we need to know more about the minimal biopolymer length needed to start life. If the minimal biopolymer length is much larger than 100 monomers then, in the random polymer assembly model, the average time for life to start in the universe is much longer than its 14 billion year lifetime and we are in this third case. In this third case, we are very likely 'alone'. It should be emphasized that highly improbable events occur in nature without the involvement of any kind of supernatural or superscientific phenomena. So this third possibility does not require any kind of 'intelligent design', 'miracle' or other superscientific intervention in natural processes. In this third scenario, no 'Fermi paradox' occurs because the failure to observe any life or civilizations elsewhere is consistent with its predictions. To get a quantitative picture of what alternative 3 would mean, we plotted the Poisson probability of life initiation (1 event) as a function of t/τ in Fig. 4.11. If $t/\tau \ll 1$ then the probability is just t/τ so, for example if $\tau = 10^{109} years$ then the event on earth occurs with a probability of about 10^{-100}. Yes, this is exceedingly unlikely, but with only one event in the sample we cannot exclude it. It would mean that, given our existence, the probability of seeing another such event is equally small, hence negligible. Therefore in this third case, searches for extraterrestrial life are exceedingly unlikely to succeed.

If alternative 3 holds, then it has some implications for the idea embodied in the so-called 'anthropic principle' [37]. There are several forms of this idea, but I will only address the 'weak' version which says that the laws of nature and the present state of the universe must be such that they allow the existence of ourselves as observers. This is trivially obvious, but many authors have emphasized that small changes in the laws of nature would make our existence (at least as we are presently consti-tuted) impossible. Alternative 3 is consistent with such statements, but would mean that though the laws of nature and the present state of the universe allow our exis-tence, our existence is extremely improbable in the present state and with the present laws. That is, if we could imagine 'running the universe again' with the same laws, we would be extremely unlikely to appear. If that is the way that the weak anthropic principle is satisfied, then it is a very weak principle indeed.

Notice that alternative 2 is not necessarily inconsistent with alternative 3, if the statements are interpreted using different definitions of 'life'. Suppose in alternative

Fig. 4.11 Poisson probability as a function of t/τ

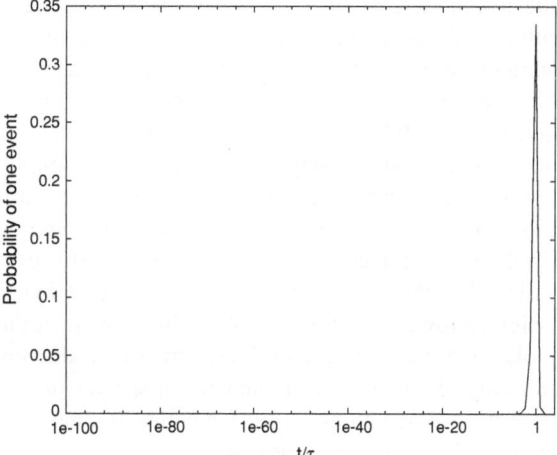

3, we are using a constrained definition of life with a hydrocarbon genome, individuals in distinct species, etc whereas in 2 we consider any long-lived metastable autocatalytic molecular system of sufficient complexity to be a 'biosphere'. Then our biosphere could be rare by the definition used in 3 but common by the definition used in 2.

4.6 The Probability f_{evol} That a Biosphere Evolves to the Complexity of a 'Civilization'

Finally we consider the factor f_{evol}. We take this to be the probability that an initial bacteria-like life system of prokaryotic, single celled organisms evolves to a biosphere of complexity comparable to the one observed on earth, capable of producing complex intentional electromagnetic signals observable from other stars. There are several unknown factors in evaluating f_{evol} and some scientists [40] have taken the view that this evolutionary step is the rate limiting slow one in the appearance of civilizations around stars. However, some data are available which at least suggest that this is not so:

The most commonly observed evolutionary process in the current practise of science and technology is the evolution of antibiotic resistant bacteria in humans and other complex organisms. This evolution, which proceeds much as Darwin originally conceived the process, is very rapid and is a serious public health problem. For a rough quantitative model, one may suppose that the rate of successful mutation is of the form $(1/\tau) \times P_{success}$ where $1/\tau$, the 'attempt rate', is the inverse of the species' reproduction time. For bacteria this is roughly $1\,h^{-1}$. $P_{success}$ is the probability per generation of a successful mutation or set of mutations leading to a new 'successful' species, in the case of bacteria, a bacterium resistant to the antibiotic in which the culture is living. Estimates [41] of $P_{success}$ for E Coli are around 10^{-9}. Inserting these numbers into the rate equation one gets a rate per bacterium of about $2 \times 10^{-5} yr^{-1}$. The 'bacterial load' per patient is reported to be [42] about 10^{10} which would then give a successful mutation to an antibiotic resistant bacterium in each patient about once a minute. This is may be a bit high but is of roughly the right order of magnitude.

Now we attempt to extrapolate to much longer times and more complex species and assume that the factor $P_{success}$ is of the same order of magnitude. It involves similar biochemical processes in all species. However the reproduction rate clearly varies. We take an average reproduction rate of 1 yr for all species. (This should be weighted toward the low end of the lifetime spectrum because there are a lot more prokaryotes than eukaryotes. The average reproduction rate could be calculated more carefully.) Estimates of the number of species on earth are around 10^7. If this has been the average number over the entire lifetime of the biosphere then we estimate the rate of appearance of new species at $10^7 \times 10^{-9}/(1\ yr) \approx 10^{-2}\ yr^{-1}$ or about 1 per century. (This is an estimate of the average. The paleontological record shows that the appearance of new species often occurs in short temporal bursts.) Multiplying the estimated average rate by the lifetime of the biosphere ($3 \times 10^9\ yr$), we would expect 3×10^7 species which is consistent with the reported numbers based on species counts. Thus a rough quantitative picture can be drawn, consistent with known empirical facts, which suggests that $f_{evol} \approx 1$. (The fact that transitional 'links' between species are rarely observed in the paleontological record also is consistent with the general picture which predicts a very rapid passage from one species to another once a critical 'saddle point' in the survivability 'landscape' has been reached.) In any case, nothing in the known empirical data base suggests tiny numbers like those found for f_{prebio}.

Some workers have argued that, though evolution always occurs once life has begun, the rate of speciation is very sensitive to the amount of variation in the environment. In the example just cited of bacteria in a hospital, the microorganisms are subjected to the huge environmental challenge of new antibiotics and rapidly evolve new dominant strains in response. In the absence of such environmental challenges, such evolution is believed not to occur. There is evidence for this idea in the fact that microrganisms dominated the earth's atmosphere for the first 2 billion years of its existence before eukaryotes and, about a billion years later, multicellular organisms appeared quite suddenly. These facts are interpreted as evidence that, when they are not environmentally challenged, species are stable and new ones do not evolve. This is the essense of the evolutionary theory of 'punctuated equilibrium'. Such a scenario would suggest that, even if prebiotic evolution proceeded rapidly, with $f_{prebio} \approx 1$, evolution to multicellularity and complexity like that in our biosphere would be very rare because many planets may have stable environments. This is the argument of Ref. [40]. In Ref. [40] the authors assume that $f_{prebio} \approx 1$ as well, so that they conclude that microorganic biospheres are common whereas complex, civilization like biospheres are rare. It may be that BOTH, f_{prebio} is $<< 1$ AND that subsequent evolution is slow because many planets have stable environments. However the stability of the environments of exoplanets is not known empirically. The environments of planets in our solar system are better known, and many of them are much more stable than that of earth, which is particularly disrupted by the plate tectonics which cause continents to collide leading to frequent earthquakes, volcanoes and rearrangements of land patterns. Similar effects only occur on a few other planets and satellites in our solar system, as we will discuss in more detail in Chap. 8. However the fraction of solar system planets and satellites which have disruptive environments is of the order of $10^{-1}-10^{-2}$ of the total. Assuming that this is typical of other planetary systems, and accepting the arguments for the need for a disruptive environment for rapid evolution would then imply a value of the same order ($10^{-1}-10^{-2}$) for f_{evol}.

References

1. L.J. Henderson, The Fitness of the Environment (1913, 1924, 1927, 1958, 1966, 1970, 1987), 1st edn. (Macmillan, New York)
2. Frank Lee, http://www.web-books.com/MoBio/ Publisher: Web Books Publishing (located in Los Angeles, USA, 2011)
3. Figure By Richard Wheeler, used with permission(http://commons.wikimedia.org/wiki/File: DB.png)
4. Figure by Deborah Spurlock, Indiana University Southeast, by permission
5. S.L. Miller, Science **117**, 528 (1953)
6. H.J. Cleaves, J.H. Chalmers, A. Lazcano, S.L. Miller, J.L. Bada, Orig Life Evol Biosph **38**, 105–115 (2008)
7. M.M. Yusupov et al., Science **292**, 883 (2001)
8. Reprinted from C.S. Tung, K.Y. Sanbonmatsu, Biophys. J. **87**, 2714–2722 (2004) Copyright 2004 with permission from Elsevier
9. I have coined the name. Sometimes the naive model discussed here is called the 'naked gene' model. See S. Rasmussen, L. Chen, M. Nillson, S. Abe, Artif. Life **9**, 269 (2003) for example. This paper also contains a useful review of 'bottom up' origin of life models.
10. M. Eigen, Die Naturwiss. **58**, 465 (1971)
11. F.H.C. Crick, S. Brenner, A. Klug, G. Pieczenik, Origins Life **7**, 389 (1976)
12. M. Eigen, *Steps Toward Life* (Oxford University Press, Oxford, 1992)
13. T. Fenchel, *Origin and Early Evolution of Life* (Oxford University Press, Oxford, 2002) Chapter 5
14. A.G. Cairns-Smith, A.G. Hall, M.J. Russel, Origins Life Evol. Biosphere **22**, 161 (1992)
15. K.I. Zamaraev1, V.N. Romannikov1, R.I. Salganik, W.A. Wlassoff, V.V. Khramtsov, Origins Life Evol. Biosphere. **27**, 325 (1997)
16. M. Eigen, P. Schuster, *The Hypercycle* (Springer, Berlin, 1979)
17. F. Dyson, Origins of Life, 2nd edn. (Cambridge Press, Cambridge, 1999); J. Mol. Evol.**18**, 344 (1982)
18. see for example H. Morowitz, D. Deamer, B. Heinz, Origins Life Evol. Biosphere **18**, 281 (1988) for discussion of spontaneous cell formation
19. S. Kauffmann, *At Home in the Universe* (Oxford University Press, Oxford, 1995)
20. S.A. Kauffman, *The Origins of Order* (Oxford University Press, Oxford, 1993)
21. R.J. Field, M. Burger (eds), *Oscillations and Traveling Waves in Chemical Systems* (Wiley, New York, 1985)
22. U. Alon, An Introduction to Systems Biology: Design Principles of Biological Circuits, Chapman & Hall/CRC, Boca Raton, FL (2007)
23. G. Wachterhauser, J. Theor. Biol. **187**, 483–494 (1997)

24. for a readable pedagogical review: http://www.cogs.indiana.edu/Publications/techreps2000/241/241.html
25. By Narayanese, WikiUserPedia, YassineMrabet, TotoBaggins [GFDL (http://www.gnu.org/copyleft/fdl.html) or CC-BY-SA-3.0 (http://www.creativecommons.org/licenses/by-sa/3.0)], via Wikimedia Commons
26. J. Cohen, I. Stewart, *What Does a Martian Look Like?: The Science of Extraterrestrial Life* (Wiley, New York, 2002)
27. G. Feinberg, R. Shapiro, *Life Beyond Earth. The Intelligent Earthling's Guide to the Universe.* (Willam Morrow, New York, 1980) and R. Shapiro, G. Feinberg, Extraterrestrials, Where are They, ed. by B. Zuckerman, M.H. Hart, 2nd edn. (Campbridge Press, Campbridge, 1995), p. 165
28. reviewed briefly in D. Schulze-Makuch, L. Irwin, *Life in the Universe.* (Springer, Berlin, 2004) Section 2.2
29. C. Fraser, E.J. Alm, M.F. Polz, B.G. Spratt, W.P. Hanage, Science **323**, 741 (2009)
30. B. Holldolber, E.O. Wilson, *The Superorganism: The Beauty Elegance and Strangeness of Insect Societies* (W.W. Norton, New York, 2009)
31. L. Margulis, *Origin of Eukaryotic Cells* (Yale University Press, London, 1970)
32. L. Margulis, *Symbiotic Planet: A New Look at Evolution* (Weidenfeld and Nicolson, London, 1998)
33. J. Lovelock, *The Revenge of Gaia: Why the Earth Is Fighting Back - and How We Can Still Save Humanity* (Allen Lane, Santa Barbara, CA, 2006)
34. J.W. Kirchner, J. Clim. Change **52**, 391 (2002)
35. M. Wong, I. de Pater (University of California, Berkeley)
36. B. Carter, Phil. Trans. R. Soc. A **310**, 347 (1983)
37. J.D. Barrow, F.J. Tipler, *The Anthropic Cosmological Principle* (Clarendon Press, Oxford, 1986) pp. 557–558
38. F. Crick, *Life Itself: Its Origins and Nature* (Simon and Schuster, New York, 1981) pp. 90.
39. D. Schulze-Makuch, L. Irwin, *Life in the Universe* (Springer, Berlin, 2004) chapter 7
40. e. g. P.D. Ward, D. Brownlee, *Rare Earth: Why Complex Life is Uncommon in the Universe.* (Copernicus, New York, 1999)
41. http://www.biotech.ubc.ca/Biodiversity/AttackoftheSuperbugs
42. http://www.pnas.org/cgi/content/full/102/37/13343

Part II
Top Down: What We Learn from the Failure of Attempts to Detect Extraterrestrial Life

In this section we consider the failure of the attempts to detect extraterrestrial life directly. This failure is sometimes termed the 'Fermi paradox' though whether it is a paradox or not will require discussion. We consider unidentified flying objects (UFO's) and the reasons that scientists have for not regarding them as evidence of visits by extraterrestrial civilizations (Chap. 5), the implications of this absence of colonisers or visitors (Chap. 6), the existing electromagnetic searches for extra-terrestrial intelligence and the implications of their failure (Chap. 7) and the searches for life on solar system planets by space probes and their implications (Chap. 8).

Chapter 5
Unidentified Flying Objects

We have seen in the first section that an attempt to determine the likelihood of observing extraterrestrial civilizations using currently known science and the Drake equation does not yield a useful quantitative result, though it constrains the possibilities significantly. This is almost entirely because not enough is known about prebiotic chemistry, but at this time that is a serious difficulty. However, there is another approach to this question which uses the fact that there are no reliably confirmed observations of any extraterrestrial biospheres or civilizations. The scientific community almost universally acknowledges this. The so-called 'Fermi Paradox' arises because some of the estimates from the Drake equation discussed in Part I predict large number of biospheres and civilizations, whereas none have been observed.

However, because large portions of the general public do not accept the notion that no extraterrestrial civilizations have been observed, this first chapter in Part II will be devoted to a brief discussion of the scientific criterion for evaluating reports of sightings, abductions and the like which are reported in the sensationalist media.

5.1 Separating Scientific from Irrefutable Claims

To make clear why almost all scientists reject claims of observations of extraterrestrials, it is useful to digress to describe the criterion by which scientists, in practise, evaluate experimental observations and the theories by which they are interpreted. This account will be quite similar to the philospher Karl Popper's description of these matters [1] but I make no claim to precisely endorse or reproduce the details of that work, or of any other writer in the philosophy of science. In practise, scientists do make reference from time to time to certain of Popper's ideas, particularly that of 'falsifiability' of a theory, but they do not spend much time on the technical details of his philosophic work, nor will I. I have observed that professional philosophers often object that Popper's work has not 'solved' certain philosophical problems of interest to them concerning the nature of induction. What is relevant here is not whether that problem was solved, but whether Popper correctly describes how scientists work and how they evaluate the significance of the results. In that respect, many scientists would agree that Popper's description (which may not have been first made by him) is a usefully careful statement of existing scientific practise and attitudes.

It can certainly be said that Popper's views are much closer to the prevailing views among scientists than are those of the various types of 'deconstructionists', often associated with the political left, who influence some departments of the history and philosophy of science in the US. Roughly speaking, these writers more or less reject the claims of science to reveal 'truth' (however, interpreted) altogether and claim that in some sense scientific theories are merely social phenomena, not related to the real world. Such views are vehemently rejected by the vast majority of scientists, because they fail to account for the overwhelming evidence of scientific success in predicting both natural phenomena and the behavior of engineered devices. They will not be discussed further here [2]. Popper's description and consensus scientific practise are even more strongly at variance with certain descriptions of 'sound science' by writers, frequently from the political right, who demand absolute proof of scientific statements by empirical evidence and who, on the other hand, often work from preconceived a priori hypotheses which are not subjected to empirical test.

Generally scientists distinguish between observations and the hypotheses which are used to interpret them. (In daily work one refers to 'experiment' and 'theory' and individual scientists, especially in chemistry and physics, often specialize in one or the other aspect, while learning a good deal about the other one.) We describe first the criterion for evaluating the validity of observations. The criterion is that any observation which is to be regarded as a scientific fact must be reproducible by anyone (not just by the individual who first reported it). Thus, in scientific reports of any observation, editors require that the authors describe their experiments so that any trained scientist could, by reading the paper, carry out an identical experiment or observation and, by observing the result him or herself, check the validity of the first reported observation. Not all reported experiments end up being checked in this way, but a great many of them, including all the ones of major significance, do. Furthermore, failures of reproducibility and rejections of initially reported results, even of quite major significance, are not unusual in scientific practise. For example, some quite recent reports of remarkable observations of single molecule electronic devices reported by workers at Bell Laboratories could not be reproduced by many laboratories and led to a very high profile investigation which resulted in repudiation of the original work. Similarly in the late 1980s reports of 'cold fusion' of isotopes of hydrogen by electrochemical methods were rejected [3] by the community because they could not be reproduced in other laboratories. In summary an observation is only regarded as a 'scientific fact' after it has been confirmed by repeated observation. In principle, this process of confirmation of observations by repeated experiments is not influenced by the theories which are used to interpret the experiments. In practice, this separation of observation from theory is usually followed quite well, and though individuals may stray from it, experiments by others with different theoretical ideas usually removes much theoretical bias from the body of reported scientific fact. Note also, that scientific facts are established if an overwhelming majority (not necessarily 100%) of the observations are consistent. The possibility of various sorts of human and other random error in carrying out experiments makes it necessary to allow for some observations which are inconsistent with the others if any facts at all are to be established.

Given a body of scientific facts, determined by this process of repeated observations, scientists attempt to use the results to test hypotheses, also called theories. The usefulness of theory is that, unlike the raw scientific facts, it can be used to predict what will happen in new experiments or observations. (Theories are often, but not always, in mathematical form so that they predict actual numerical values for the results of the new experiments or observations.) The relation of theories to the 'real world' glimpsed by scientific facts is that theories are regarded as provisionally true until proven false. It is a very important feature that no statement or set of statements (in mathematical form or not) can be claimed to be a scientific theory unless it can be used to rigorously deduce predictions about the results of experiments or observations which can then be carried out as a test of the theory. Notice that the theories must predict the results of experiments or observations to be carried out in the future, but the events generating the data accumulated in doing the experiments or observations may have occurred in the distant past. For example the theory that the extinction of the dinosaurs at the Cretaceous-Tertiary boundary approximately 65 million years ago was the result of a meteoric impact generated the prediction that remanents of the resulting crater could be found. The later observation of the remanents of such a crater in Yucatan, Mexico was regarded as a success of the theory.

It must be possible for the results of the experiments or observations to be other than those predicted by the theory. This is the idea that theories must be 'falsifiable'. This means that the body of scientific theories is never to be regarded as proven once and for all. All scientific theories are ever subject to refutation by new experiments. On the other hand, some scientific theories are rightly regarded as being on much stronger ground than others, because they have been tested for centuries by new experiments [4]. The way in which successful tests can be used to evaluate the relative probability that different theories which are consistent with them are correct has been worked out (Appendix 5.1) though scientists do not very frequently use this methodology. In many cases, the tests eventually find some phenomena for which the theory gives false predictions. In such cases the theory might then be rejected altogether, but more commonly it is found that its domain of validity can be rationally restricted so that one can continue to use it in an appropriate domain to make predictions. For example, Newton's laws of gravitation passed all astronomical and terrestrial tests devised for them for more than two centuries. An example of an early and very influential test was the prediction by Edmund Halley, using Newton's theory, of the time of return of the comet which is now named for Halley. However, in the early twentieth century, it was found that certain features of the orbit of Mercury did not conform to Newtonian predictions and the fundamental theory of gravity was replaced by the theory due to Einstein called general relativity. Nevertheless, Newton's laws are still used to successfully predict many phenomena (including, with astounding accuracy, the trajectories of space craft) because the range of gravitational fields and velocities for which they are valid has been determined.

Another criterion for good scientific hypotheses, also described by Popper [5] but originally attributed to William of Ockham, a fourteenth century logician, is that hypotheses should be a simple as possible while still agreeing with the known facts. From the point of view of falsifiability, simple theories are preferrable because they are easier to test.

In the context of our subject it is useful to consider one simple generic example of a theory. Consider two proposed theoretical statements

1. A exists.
2. A does not exist.

and suppose that A has not been observed in a repeatable way as discussed above at the present time. We also assume that we have determined well defined experimental procedures for recognising A if we observed it. This is not a trivial point in our case. Now 1. is not an easily falsifiable theory, since no failure to observe A can be said to refute 1. (It can simply be said that one did not look in the right place at the right time.) However, any observation of A will refute 2. In fact, though 1. is not easily falsifiable, one might claim that it could be falsified by looking everywhere simultaneously. But this is clearly highly impractical if not impossible. Obviously we are thinking of A as extraterrestrial civilization. From a scientific point of view, the hypothesis that extraterrestrial civilization does not exist is the preferable one, because it can be refuted by just one observation of an extraterrestrial civilization. This is what the SETI searches to be described in Chap. 7 are trying to do.

Now turning to our subject, how are we to evaluate reports of unidentified flying object (UFO) sightings? A recent review of evidence appears in [6]. (Though this reference concludes that UFO reports do not provide evidence for extraterrestrial life, some have suggested that the panel which produced the report was more open-minded than is warranted. See reference [7] for example. The entire subject of UFO reports has been the subject of controversy for more than 50 years. Some of the early history is summarized in reference [8].) There are thousands of reports. Almost all of them consist of visual reports by witnesses of transient phenomena that do not seem to recur at the same place or time of day. Many of them have possible explanations in terms meteorological or optical phenomena. Some of the accounts definitely suggest psychiatric explanations involving the mental processes of the witnesses. There is a residue of unexplained reports of various sorts, some involving physical alterations of the ground, radar anomalies, malfunctioning vehicles and injuries to witnesses. There is almost no quantitative data (Photographs and video tapes are in the main fuzzy and uninterpretable.)

It should be evident that there are few, if any, scientific facts in the sense described above, which are established by this information. Thus the first criterion for making and testing a scientific hypothesis, namely a body of established scientific fact, is absent. This makes the evaluation of hypotheses essentially impossible. In view of the situation, and for the reasons cited above, the most appropriate scientific hypothesis concerning visits of living extraterrestrial beings to our planet is that they have not occurred. This does not mean that the hypothesis should not be tested by further observation. How much effort should go into such testing depends in part on estimates from other, non-UFO, sources concerning the likelihood of extraterrestrial civilization. Those estimates are the subject of previous and subsequent discussion in this book.

5.2 Conspiracies

In commenting on this subject of UFO reports one cannot fully avoid a very common idea that the lack of evidence is a consequence of a government conspiracy. In view of repeated revelations of supposedly secret government programs in the press, some of them involving very sensitive issues (for example, rather detailed plans for attacking Iraq in major newspapers of the US before the US invasion of Iraq), the ability of the government to keep a major secret concerning this matter for more than 6 decades may be reasonably questioned. Indeed in the specific matter of UFO reports, the US Air Force and others in the government in the late 1960s apparently did attempt [8] to influence the conclusions of the so-called Condon report [9] on the subject, but this came to light very soon. In that case, the government motivation was apparently to resist pressure for further research. In a broader context, it is hard to understand what motive the government might have for covering up these supposed facts. The Defense Department (of which branches are usually alleged to participate in the coverups) thrives in terms of its power, influence and budget when the nation has clearly identified enemies. Proof of identification of visits by extraterrestrial civilization, which could certainly at least be portrayed as potentially threatening, would seem to be very useful to the Defense Department in its quest for money and power. But the main point here is that the hypothesis of a government conspiracy is usually framed in what we define as an unscientific way: It is nonfalsifiable because no matter how thoroughly one explores government documents, uses Freedom of Information Act procedures, interviews officials, etc. it will always be claimed by proponents of these conspiracy theories that the real facts are hidden under deeper levels of secrecy by the infinitely resourceful and diabolical government. A falsifiable hypothesis of this general type would have to be more specific e.g.: 'Evidence will be found in the archives of the US Air Force that the wreckage of a vehicle from an extraterrestrial civilization was found in Roswell, New Mexico in 1948.' Such statements are falsifiable. However, based on the fact that the government very seldom succeeds in keeping secrets for a long time, a better falsifiable hypothesis is that the government has not succeeded in hiding any scientific facts of major qualitative importance regarding UFO phenomena. The latter will be falsified by any discovery of a coverup in government archives. Conspiracy theories are unlikely to be correct whatever one thinks of the ethical nature of government officials. However, the assumption that the government officials are entirely evil is as unrealistic as the assumption that they are entirely virtuous.

5.3 Hypotheses on the 'States of Mind' of Extraterrestrials

Assuming, as we will henceforth, that UFO sightings and other such reports have not established the existence of extraterrestrial civilization, one can ask why we do not have this kind of strong evidence of their existence. One class of hypothetical explanations, which we consider later, basically takes the view that extraterrestrial civilizations are so infrequently found in the universe and our searches for them are so limited, that we are unlikely to have found them yet. This class of hypotheses can be cast in a quantitative, scientific form and we will discuss them in Chaps. 6 and 7. Another class of proposed explanations involves speculations about the thought processes of the hypothetical extraterrestrial civilizations. Two hypotheses of this type which we will briefly discuss are

1. The contemplation hypothesis: The extraterrestrial civilizations are not interested in communicating or in colonizing so there are no signals to detect and no colonialists are expected to arrive in our solar system.
2. The zoo hypothesis: The extraterrestrial civilizations regard the earth as a primitive life system in development and are studying it while hiding their existence from us.

In principle both hypotheses are falsifiable because any observation of an extraterrestrial civilization will falsify them, at least in forms in which it is assumed that the putative civilizations are infinitely capable of hiding themselves. However, they certainly both violate the criterion of simplicity. It is much simpler to hypothesize that the civilizations which have not been observed do not exist. In fact these hypotheses have been constructed in a way which makes them nearly impossible to test in detail: To disprove the detailed mechanisms for their nonappearance which are postulated, one would have to find the civilizations. But if one assumes, as is usually done, that they are much more advanced than humans, then by assumption that will be nearly impossible.

Further, if we include ourselves among the civilizations to which the hypotheses applies, then these hypotheses do not fit the existing facts. Humans are not contemplative in the sense suggested in 1) and are not regarding any other planet as a kind of nature preserve. The hypotheses also assume essentially perfect coordination among and within the supposed civilizations in order that they remain hidden. But, if we use our experience of terrestrial biology as a guide, the postulate of perfect coordination among and within the supposed civilizations is completely inconsistent with human experience. And the discipline and coordination required can be expected to get more difficult to maintain as the number or size of the assumed civilizations gets larger.

There could possibly be some hints in recent human behavior, particularly with respect to addiction to virtual worlds and conscious altering drugs of various sorts, that suggest a propensity for withdrawal from engagement with physical reality which is reminiscent of the behavior postulated in the contemplation hypothesis. (Oddly, some have suggested use of virtual reality to *cure* physical addictions. However, the similarities of addiction to physical substances and various kinds of computer oriented activity [10] make it seem more likely that both kinds of activity are symptoms of a similar tendency to withdraw from aspects of physical reality.)

Reading further into assumptions made explicitly or implicitly by the authors of these hypotheses, one finds that they often assume extremely long lifetimes, of the order of millions of years, for the civilizations. We will return to this issue in the next chapter, but comment here that, on the basis of human history, one would have to judge this extremely unlikely. Human written history has lasted only 10^4 years and its 'electromagnetic lifetime' (the time during which detectable electromagnetic signals have been emitted) is only about 10^2 years. Yet humans have come perceptibly close to self annihilation several times.

References

1. K.A. Popper, *The Logic of Scientific Discovery*, 2nd Harper and Row Torchbook edn. (Harper & Row, NY, 1968)
2. readers interested in the contemporary discussion of these issues may be interested in D. Miller, Pli 9, (2000) pp. 156–173 and references therein.
3. http://www.ncas.org/erab/sec1.htm
4. At least that is the view of the overwhelming majority of scientists and engineers. Some philosophers interpret Popper to mean that a theory which has passed repeated tests is not more likely to be right or useful than one which has not passed many tests. But my (limited) reading of Popper is not consistent with that interpretation and, in any case, no scientist would agree with it. See also Appendix 5.1 which describes a way to use the results of experiments to evaluate the likelihood that a theory is correct.
5. K. Popper, 7. *Simplicity. The Logic of Scientific Discovery*, 2nd edn. (Routledge, London, 1992) pp. 121–132
6. P.A. Sturrock et al., Physical evidence related to UFO reports. J. Sci. Explor. **12**(2), 179 (1998)
7. http://www.tampabayskeptics.org/v11n3rpt.html#SSE
8. http://en.wikipedia.org/wiki/Condon Committee
9. available on line as http://ncas.org/condon/
10. H. Phillips, New Scientist, 26 Aug 2006

Chapter 6
Colonization and Panspermia

UFO reports cannot be regarded as establishing that extraterrestrials have visited the earth. On the other hand, if other biospheres exist in the galaxy, it is conceivable that life forms might migrate from one star to another even though the distances between stars are more immense than is realized by most people. These two circumstances lead to the following argument [1], which people have attempted to use to put some bounds on the numbers and types of biospheres which are likely to exist in the galaxy. In rough qualitative form, the argument is

1. Extraterrestrial visitors have not appeared on earth during human history.
2. This presumed fact (1) is evidence, not that interstellar travel is impossible, but rather that extraterrestrial civilizations do not exist in the galaxy at this time.

We speak somewhat loosely about 'this time' here. We really mean 'have not existed within a past time long enough for them to have gotten here from elsewhere in the galaxy'. Since in most scenarios this time would not be longer than 10^7 years or so, this difference will not make a large difference in the conclusions.

A similar discussion is possible if we are not interested in 'civilizations' but in any form of life at all. It is less clear in that case that no form of life has appeared on the earth from elsewhere, though the biochemical unity of the biosphere already discussed would suggest that if 'seeding' from elsewhere occurred, it probably only occurred once. The forms of interstellar transport possible for very primitive forms of life are probably less restrictive than those expected to be possible for complex forms (though this is not obvious and could be wrong) so we will make a separate discussion of the merits of the argument for this second case.

6.1 Implications of the Failure to Observe Colonization of Earth by 'Advanced' Civilizations

Because misconceptions about the time and length scales involved in speculations about interstellar travel are widespread, we will begin with a discussion of what would be possible in principle for humans with existing technology. It is represented

J. W. Halley, *How Likely is Extraterrestrial Life?*, SpringerBriefs in Astronomy, DOI: 10.1007/978-3-642-22754-7_6, © The Author(s) 2012

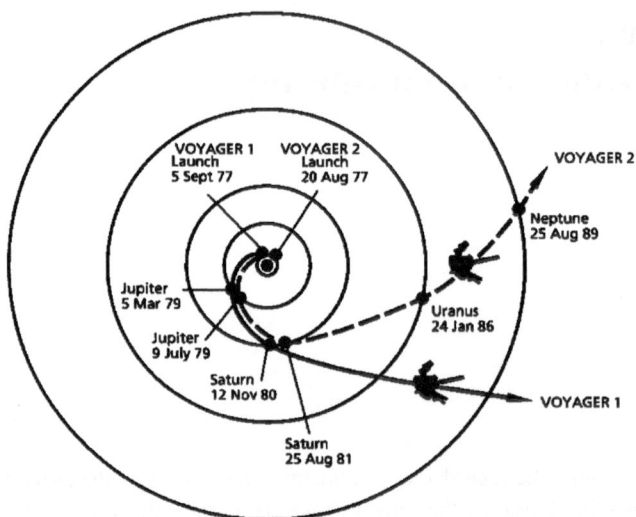

Fig. 6.1 Path of the Voyager spacecraft through the solar system

by the Voyager missions, which are two unmanned spacecraft launched in 1977 from the US. After a tour of the outer planets, both space craft proceeded out of the solar system and , in 2008, were approximately 1.5×10^{13} m from the earth and sun traveling at about 16,800 m/s (34,000 mi/hr) away from earth and the sun (Figs. 6.1 and 6.2.) They are still in radio contact with earth and are transmitting data about the interstellar medium. The nearest star (Proxima Centauri) to our sun is about 4 light years $\approx 3.6 \times 10^{16}$ m away. Thus these space craft would require about 66,000 years to reach it. Actually neither craft is going toward Proxima Centauri. In about 40,000 years, Voyager 1 will drift within 1.6 light years (9.3 trillion miles) of AC+79 3888, a star in the constellation of Camelopardalis. In some 296,000 years, Voyager 2 will pass Sirius, the brightest star in our sky, at a distance of about 4.3 light years (25 trillion miles). (See the Voyager web site [2], for current information about the Voyagers.)

In discussions of space travel, it is usually assumed that these long times to reach neighboring stars with current technology make interstellar space travel totally impractical unless means can be found to greatly increase travel speeds. The difficulty is that , as the speed increases to near the speed of light in vacuum (3×10^8 m/s) the energy cost (which is roughly proportional to the cost in resources and in payload) does not increase as the square of the speed v, as it does at low speeds, but as $1/\sqrt{(1-(v/c)^2)}$ which enormously increases the projected costs of accelerating any vehicle to a significant fraction of the speed of light. Detailed studies [3], using technologies which are extensions of existing ones, estimate costs on earth of the order of $\$ 10^{12}$ (1985 US) for a manned trip to a nearby star which could be launched within a time between 10^2 and 10^4 years from the present [4]. with ship speeds of order $10^{-1} c$. There is a serious question about what motive humans would have to do this

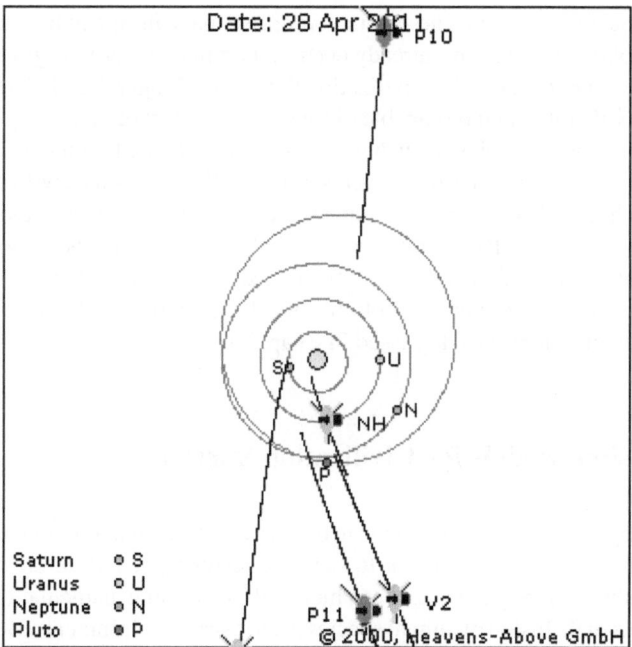

Fig. 6.2 Positions of four spacecraft leaving the solar system as of spring 2011. See Ref. [5] for regular updates. The labels refer to the Pioneer 10 (P10), Pioneer 11 (P11), Voyager 2 (V2). The position of Voyager 1 is just off the figure at the end of the arrow at bottom left. NH stands for "New Horizons" , which is a satellite enroute to Pluto

in such a short time. There appear to be much easier ways to deal with overpopulation than interstellar travel. Over times of order 10^{10} yrs sunlike stars become red giants and earthlike planets at 1 AU from the star would become uninhabitable. However even optimistic scenarios do not envision civilization lifetimes of 10^{10} years.

The presumed need for humans to reach speeds of order 10^{-1} c in order to make interstellar travel practical is based implicitly on the assumption that human life expectancies will stay below 10^2 years. If human longevity were to remain of this order, trips even to nearby stars with present spacecraft would require more than 100 generations and no conceivable human society would support such voyages. However it may be a mistake to assume that human longevity will not change over the very long civilization lifetimes assumed by many studies of the likelihood of interstellar travel (several writers assume millions of years). Indeed we have no reason to suppose that it is so short now in a hypothesized extraterrestrial civilization. Human life expectancies have increased by around a factor of three in the last 2,000 years and there are some indications that the rate of increase may be increasing. If longevity were 10^4 years instead of 10^2 years then interstellar voyages, even at the presently feasible 1/10,000 of the speed of light, would be more imaginable. One should also note that these discussions presume that the interstellar voyages carry members of the species of the extraterrestrial civilization.(These points were also made in Appendix A of Ref. [3].) Nonliving robotic probes have already been sent on interstellar voyages by humans.

It may be concluded that if a civilization lives much longer than 10^4 years, then it is quite likely that it would be capable of sending some sort of machine possibly carrying members of its species (if such a concept can be defined for the civilization) to nearby stars with travel times of 10^4 years or less. Whether such a civilization would do so is much less clear given our estimates based on attempts to extrapolate human experience. (The idea that extraterrestrial civilizations would have no motive to colonise is one aspect of what is sometimes called the "Contemplation Hypothesis", which holds that civilizations are not observed because they lack the motivation to communicate or colonise as discussed in Chap. 5.)

6.2 Diffusion Models for Civilization Spread

Now we suppose hypothetically that such motivations exist and the presumed civilization begins to migrate from star to star. The resulting pattern of diffusion of the civilization through the galaxy has been modeled [6, 7], using mathematics quite similar to models of diffusion of animal species in terrestrial environments. Because such models are rather complicated and require uncontrolled assumptions at the outset we will not discuss them in detail. General features of the results can be understood from a simple diffusion picture (though this picture gives quite a different shape for the diffusion front than the nonlinear models do.) The basic idea of diffusion is quite simple (see Appendix 6.1 for some more details). To describe the motion of a collection of bodies through space, one divides time into relatively small intervals which last a time τ_{hop} and assumes that each body chooses a random direction at each interval of time and moves a fixed step length (which we will call l) in that direction. The model is easily generalized to allow for a distibution of different step sizes but the results, within some rather broad set of distributions, are the same as for the model just described. It is quite easy to show (Appendix 6.1) that in this model the average distance which a body moves from its starting point after time t is $l\sqrt{t/\tau_{hop}}$. This is often written as $\sqrt{6Dt}$ in three dimensions where D, the diffusion constant is $D = l^2/6\tau_{hop}$.

If simple diffusion is an appropriate model, then the diffusion constant would be of order $l^2/6\tau_{hop}$ where $l \approx R/N_{gal}^{1/3}$ is of the order of the distance between stars and τ_{hop} is the average hop time which we have argued is probably 10^4 years or less. Then the time for the random walk path followed by the civilization to fill the galaxy is given by $N_{gal}^{2/3}\tau_{hop}$ which is 10^{11} yr if τ_{hop} is 10^4 years and 10^8 yr if τ_{hop} is 10 yr (which is near the shortest conceivable hop time using known physics). The nonlinear models mentioned above are reporting fill times of 10^8 yr with ship speeds of 10^{-1} c so these results are roughly consistent.

However, this estimate is quite model dependent. For example, if we use the possibly more realistic model of a self avoiding random walk [8] then the mean square

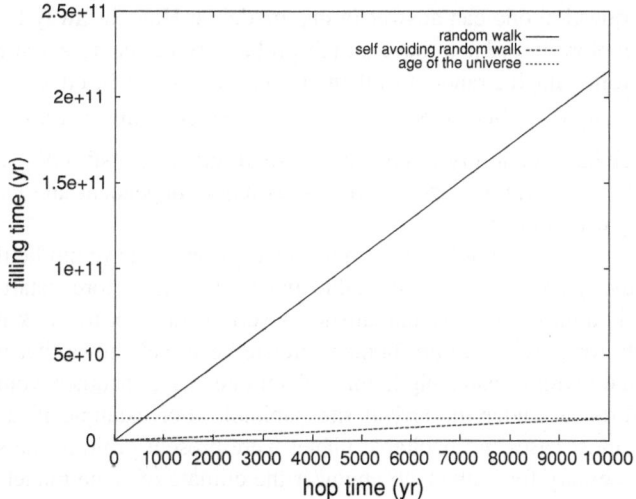

Fig. 6.3 Estimated filling times as a function of hop times in the random walk and self avoiding random walk models. The horizontal line is the age of the universe

deviation distance from the origin is approximately proportional to the number of steps to the 6/5 power (rather than the power 1 for a random walk). Then the time required for filling the galaxy becomes $N_{gal}^{5/9} \tau_{hop}$. This apparently minor difference could affect the conclusions as illustrated in Fig. 6.3. In another variant of the random walk model (called a branching random walk [9] there is a finite population at each site, and each member of the population sends an 'off-spring' to a randomly selected neighboring site. Thus, applied to our problem, civilizations could propagate from a given star to more than one star at each time step. This modification is not expected to affect filling times but it could affect the density of civilizations achieved at the filling time (as discussed below for the random walk model).

In the random walk model we thus predict that a civilization with the motivation to colonise would fill the galaxy during the age of the universe only if its technology permitted hops every few hundred years whereas in the self avoiding walk model there would be enough time to fill the galaxy even if hop times were of order 10^4 years.

The random walk diffusive model reveals another feature relevant to the argument. In the random walk model, we estimated the time for the civilization to diffuse out to a sphere with a radius of the order of the size of the galaxy. But this is not actually the 'filling time' in this model, because at that time, the walk has not visited every star. In fact one can see from the foregoing that it has visited $N_{gal}^{2/3} \approx 2.1 \times 10^7$ stars. The probability that our star was visited in the 'filling time' is therefore $N_{gal}^{2/3}/N_{gal} \approx 2 \times 10^{-4}$. The numbers are smaller for the self avoiding walk model. Of course the civilization could continue exploring the galaxy after it reaches the edge and the random walk model is certainly oversimplified, but even in more sophisticated models, it may be that simple estimates of 'front speed' could overestimate the probability of being visited.

Another question one can answer in this model is: How far away from our star should we explore in order to have a high probability of seeing a visited star, if a civilization following the random walk model has diffused to the edge of the galaxy? Since the density of visited stars is $N_{gal}^{2/3}/(4\pi R^3/3)$ the radius r of a sphere around our star which has probability 1 of having a visited star in it satisfies $N_{gal}^{2/3}(r/R)^3 = 1$ which gives $r \approx 35$ light years. Again, this is model dependent and larger for the self avoiding walk model.

In putting an upper bound on τ_{hop} in this discussion, we have implicitly assumed that the diffusing entity spends a negligible time at each star before it starts migrating again. This is almost certainly unrealistic. A serious attempt to think through the requirements for possible future human interstellar travel shows that most of the time associated with transferring humans from one star to another would be taken up with efforts to render the arrival site habitable. For example if, as would be expected in most cases, no habitable planet were present orbiting the star, then it would be necessary for humans to engineer the climate of some planet which was present to render it habitable [10] (this idea is called 'terraforming') or to build artificial dwellings orbiting the star. Either effort can easily be imagined (and has been estimated) to take orders of magnitude longer than the actual interstellar trip. Thus the upper bound on τ_{hop}, which was based on present human capabilities, is probably too low. We take this into account in the conclusions later.

In either of the two models considered, the filling times are long. The shortest time, obtained with $\tau_{hop} = 10\,yr$ and the self avoiding walk model, is about 10^7 yrs. In fact, a more complete description should include the birth and death rate of civilizations. In such a model, the steady state density of civilizations will be determined by the sum of local birth rate, local death rate and immigration – emigration rate:

$$-n_{civ}/\tau_d + n_{gal}/\tau_b + diffusion = 0$$

where the diffusion term describes the difference between local immigration and emigration. Here n_{civ} and n_{gal} are local densities of civilizations and of stars. (See the Appendix 6.2 for some mathematical details and a more mathematically rigorous solution to this problem.) As discussed, the diffusion term is of order $-(l/R)^2 n_{civ}/\tau_{hop} = -(1/(N_{gal}^{2/3}\tau_{hop})n_{civ} = -n_{civ}/\tau_{fill}$ where τ_{fill} is the previously estimated filling time. Thus the local density of civilizations in steady state is of order $n_{civ} \approx (n_{gal}/\tau_b)(1/(1/\tau_d + 1/\tau_{fill}))$. So the local concentration of civilizations, which will determine how likely we are to encounter one, is determined by the death rate if $\tau_d << \tau_{fill}$ and by the diffusion rate if $\tau_d >> \tau_{fill}$. Our estimated filling times were 10 million years or longer. Thus if civilization lifetimes are less, on average, than 10 million years, then diffusion, and hence colonization, is irrelevant to the question of how likely we are to encounter an extraterrestrical civilization. (The somewhat more rigorous discussion in (Appendix 6.2) gives a similar lower limit.)In that case, the local density of civilizations is just $n_{civ} \approx n_{gal}(\tau_d/\tau_b)$ which is a form of the Drake equation. In the other case in which $\tau_d >> \tau_{fill}$ we are dealing with very long lived civilizations and the local density $n_{civ} \approx n_{gal}(\tau_{fill}/\tau_b)$ is determined by the diffusive filling time and the most likely way to encounter another civilization is by colonization. Thus if our bounds on filling times are approximately right, then the argument that the failure to observe colonization by other civilizations is evidence that they do not exist only applies to civilizations that live a million years or longer. The argument is only mathematically rigorous in a simple diffusion model for migration, though one may argue that some of the qualitative features may survive in more sophisticated models.

The proponents of the argument outlined at the beginning of this chapter argue that the times for a civilization to fill the galaxy by colonization are so short that the absence of colonisers in the earth's neighborhood is evidence that civilizations of the postulated sort do not exist. It appears from our discussion that this statement may be defensible if it is sharpened along the following lines: The absence of colonisers is evidence for the nonexistence of civilizations in the galaxy which (1) live much longer than 10^6 yr (2) have achieved technology permitting relativistic interstellar travel and very fast acclimatization to stars at which they arrive and (3) are motivated to colonize. (The condition 3) excludes civilizations like those described in the contemplation hypothesis. However our analysis suggests that civilizations that were otherwise quite aggressive and extroverted might not choose interstellar colonization.) Entities which did not achieve fast acclimatisation would have to live much longer to fill the galaxy and the required times would not allow filling at all if the resulting hop times were much more than 10^5 years.

6.3 Estimating Civilization Lifetimes

The preceding shows that the argument that the failure to observe colonization implies that civilizations do not exist only applies to very long-lived civilizations. It is sometimes argued on the basis of human history that very long lifetimes are to be expected in biospheres in which very complex lifelike systems have evolved. Here I provide some arguments which cast considerable doubt on the idea that the human example provides much reason to think that civilizations are likely to survive for very long times.

Let E be the event that the present time is less than or equal to some time t in the life of our civilization [11] and H be the event that it lives a time τ_d and then dies. If we are thinking of 'electromagnetic lifetime' and if $t = 10^2$ yr then E has occurred. Consider the probabilities $P(E, H)$ that both events occur, $P(E|H)$ that E occurs given that H occurs $P(H|E)$ that H occurs given that E occurs, $P(E)$ that E occurs unconditionally and $P(H)$ that H occurs unconditionally. They are related (by a general feature of probability) by

$$P(E, H) = P(H|E)P(E) = P(E|H)P(H)$$

or

$$P(H|E) = P(E|H)P(H)/P(E)$$

Now we measure time from the beginning of our 'electromagnetic' period so our current lifetime is 10^2 yr. Assuming that the present ('now') is a randomly selected moment in time, the probability $P(E|H) = t/\tau_d = 10^2$ yr$/\tau_d$. $P(H)$ is unknown but we expect $P(H) < P(E)$ if $t = 10^2 yr$ because then E has occurred and $P(E) = 1$. Thus the probability that our civilization will live a time τ_d which is much more than the 100 years which it has already lived is bounded by

$$P(H|E) < 10^2 \ yr/\tau_d$$

and becomes very small as τ_d grows larger than 10^4 yr for example. This seems quite plausible given the everyday evidence of instability in human society. (See Ref. [11] for another form of the argument which states bounds on lifetimes more quantitatively.) The argument uses the fact of our existence as a part of the probabilistic reasoning. However it suggests that optimistic projections about the human future should not be used to estimate the average lifetime of civilizations.

One can buttress the plausibility of this argument by considering what would be required for human civilization (or some 'posthuman' successor) to survive for a million years. If ϵ is the probability per year of extinction and assuming that $\epsilon << 1$ we have the the probability of surviving N years is $(1 - \epsilon)^N$ and because $\ln(1 - \epsilon) \approx -\epsilon$ this can be written $e^{-N\epsilon}$ Thus if this is, for example, to be $>10^{-4}$ when $N = 10^6$ we would need $\epsilon < 4 \ln 10 \times 10^{-6}$/yr.

Most extinction probabilities are extremely hard to estimate (see for example Ref. [12].) However one relatively straightforward possible cause of extinction is impact of the earth by a large asteroid. Such an event caused the extinction of the dinosaurs 65 million years ago and repetition is by no means ruled out. NASA maintains a program of surveillance of near earth objects (NEOs) in collaboration with scientists from other countries and posts information about hazardous ones [13] (An image of such a near earth object appears in Fig. 6.4). For example object 2008 AF4 is given a probability of about 4.3×10^{-5} of impacting the earth during the next century (most probably in 2089) giving a contribution to ϵ of about 4.3×10^{-7}/yr from this object alone. Another object, 2007 VK184 is given a probability of 3.4×10^{-04} (in 2048) contributing 3.4×10^{-6}/yr to ϵ. These are the objects currently listed as most hazardous. It appears that the contribution of asteroid impact alone to ϵ would give a probabilitiy of extinction in 1 million years which would be large enough to account for the bound given by the argument in the last paragraph. Here I am not taking account of the possibility that humans may, in the future, actively intervene to deflect the orbits of NEO objects from collision courses with the earth. Such efforts might reduce the probability of extinction by asteroids. The efforts would be unprecedented and I do not know how to estimate the likelihood of their success. Given the enormity of the task, its totally unfamiliar aspects, the need for rather prompt action and the requirements of large scale international cooperation and very high cost, I would guess that the likelihood of success is small, at least in the next century.

This discussion of asteroid impact probabilities gives estimates of lifetimes which may be argued to be somewhat less anthropocentric than the arguments given earlier: Many, possibly most, earth like planets will be subject to similar hazards.

Fig. 6.4 The asteroid 1999 jm8, believed to be charactistic of the near earth objects which are monitored for the hazards which they represent to earth. This particular object was not deemed hazardous but is shown [14] because the availibility of this radar image shows the irregular shape thought to be typical of such objects. 1999 jm8 is about 3.5 kilometers in diameter

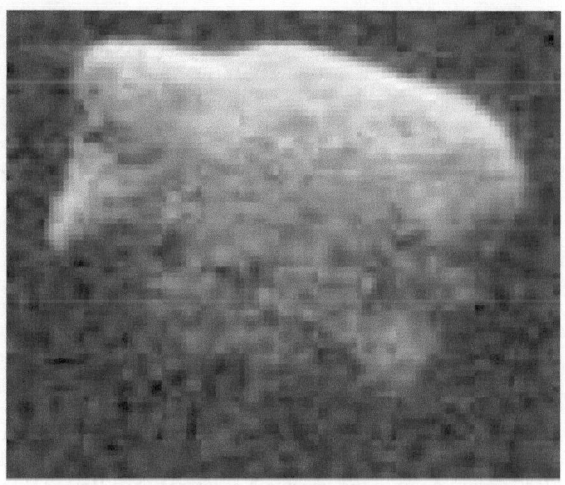

6.4 Implications of Bacterial Colonization: Panspermia

The idea that life on earth may have been seeded, or that seeding might still be occurring, is called panspermia and two forms of it have been suggested. In one version the organisms move randomly through the galaxy from some original planet [15]. In another version [16], 'directed' panspermia, the life forms have been directed, possibly at our star because it has a promising environment for life, by an advanced civilization elsewhere. There seems to be little evidence, one way or the other, concerning whether the very initial stages of life arose on earth spontaneously or were, on the other hand, imported from elsewhere in the form of primitive organisms. This uncertainty is part of our general ignorance concerning the origin of life. Such importation would not relieve the difficulties in accounting for the origin of life on the random polymer model because, wherever life might arise, it has only about 14 billion years to do so and as discussed earlier, this time is too short for most random genome assembly models.

We can be quite sure that importation of microorganisms to earth has not occurred frequently, because such importation would be expected to introduce new chemistries (at the very least different chiralities) into the biosphere and these are not observed. However, the possibility that life could migrate at least from one planet to another is open. In the case of undirected panspermia, the micro-organisms might be 'launched' into trajectories escaping the gravitational pull of their home planet as a result of impacts of incoming meteorites on the planet surface. The impacts result in recoil of massive objects which can escape from the planet, possibly carrying resident microorganisms with them. In the case of Mars and the earth, there is very strong isotopic evidence that such migration of material from Mars to the Earth has occurred [17–19]. Whether transport of such macroscopic chunks of matter occurs between stars seems to be much less well established.

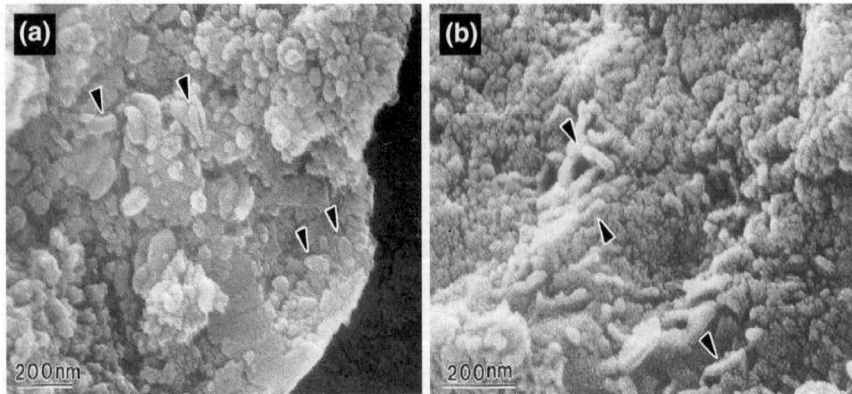

Fig. 6.5 Microscope images of objects in a meteorite, found in Antartica and shown to be likely to come from Mars, which were thought to show possible evidence of worm-like structures. The black arrows point to the structures of interest. See Ref. [22] for details. Figure from Ref. [22] by permission

For the case of transport from earth to Mars, there was a serious suggestion in 1996 that some of the meteorites whose provenance was almost certainly Martian contained fossils of microorganisms [20]. An image which was used to buttress this claim is shown in Fig. 6.5. Unfortunately, it proved possible to reproduce the possibly worm-like structures observed in the images using a crystal growing technique involving no biological components [21]. As a result of this and other studies, the current consensus is that the meteorites studied in Ref. [21] probably did not contain fossils of microorganisms though the group originally proposing the biological interpretation of the data continues to vigorously defend it [22]. Nevertheless, it is generally agreed that such transport between planets and maybe also between stars would be possible.

Whether bacteria aboard such a meteorite would survive the trip is problematical. There has been considerable excitement in the astrobiology community about the discovery of radiation resistant bacteria. However it seems that no 'extremophile' bacteria have been found which are simultaneously resistant to the extremes of radiation, of temperature and of dessication which are likely to be present during the voyage [23]. On the other hand, if the radiation dose in space is of the order of [24] 3rem/yr then the dosage a bacterium would receive in a 10,000 yr trip between stars would be 30,000 rem while the most radiation resistant bacteria can tolerate dosages of $3x10^6$rad $> 3x10^6$ rem (because quality factors are always >1. See the Appendix 6.3 for a review of radiation units.) Thus in terms of radiation hardiness alone, the radiation resistant bacteria could probably survive an interstellar trip.

With regard to 'hop times' for bacteria randomly launched from planets, the limits which were derived earlier for civilization hop times involved the speed of light and interstellar distances and therefore the lower limit would be similar at about 10 years. However 10 years seems rather unlikely and one might expect the range to shift toward longer times for the case of undirected panspermia. If hop times are as big or larger than 10,000 years then we get diffusion filling times which are much longer than the age of the universe (see Fig. 6.3. Simple diffusion is much more likely to be a good model for undirected panspermia than for civilization transport because bacteria are unlikely to be able to control where they go and are therefore expected to be transported randomly.). As in the case of civilization diffusion, diffusion of bacteria through the galaxy would only affect their local concentration if the lifetime of the bacterial colonies were much longer than the filling times. Though bacteria are essentially 'immortal' under favorable conditions, they cannot live longer than the age of the universe. We have just argued that expected filling times are much longer than the age of the universe. Therefore undirected panspermia is irrelevant to the determination of the local likelihood of finding life except in the unlikely case of relativistically short bacterial hopping times.

Directed panspermia is harder to dismiss. It is suggested [16] that an advanced civilization facing death due to some event such as the expiration of its star as a supernova or red giant might choose to launch a directed, not randomly moving, probe loaded with the essential elements of its biochemistry toward a hospitable star. In this case it might be possible to approach the 10 year hopping time limit imposed by relativity and and one might argue that diffusive and other stochastic models for transport are irrelevant. From the 'top down' observational point of view there are some 'Fermi paradox' types of arguments to the effect that this has not taken place, for example: 1) if it occurred often, then we would expect to see a variety of unrelated lifeforms appearing on earth, as we do not and 2) it requires the existence of very sophisticated other civilizations, which , by other arguments which we have and will discuss, may be unlikely to exist. However it does not appear to be easy to falsify the hypothesis of directed panspermia which may therefore be judged unscientific by a Popper like criterion. Most proposed tests are not conclusive if one postulates a sufficiently advanced and powerful extraterrestrial civilization.

6.5 Conclusions from the Failure to Observe Evidence of Colonization

In summary, our analysis of the argument [1] that the failure to observe colonization implies that other civilizations don't exist does strongly suggest that it is valid for civilizations that live longer than a few million years, are able to acclimatise quickly to new stars and are motivated to keep moving from star to star promptly after acclimitization. We can therefore conclude that such civilizations are unlikely to exist. However the list of conditions on this statement is long and the argument does not appear to have anything to say about civilizations which do not meet all the conditions. Thus the argument for nonexistence from the failure to observe colonisers only excludes a special class of conjectured civilizations.

The analogous argument applied to 'panspermia' or the transport of microorganisms through space indicates that such transport cannot be dismissed for different reasons in its two formulations. In the case of undirected panspermia, unless the meteorites carrying microorganisms between stars move at relativistic speeds, they cannot significantly affect the likelihood of finding life on a planet. Such random transport at lower speeds does appear to be possible however, and we cannot exclude the possibility that at least interplanetary transport might account for the origin of life on earth. This possibility is not of much help in resolving the question of the ultimate prebiotic origin of life, because it does not affect the 13.8 billion year time limit imposed by the age of the universe. Directed panspermia does not appear to be a falsifiable hypothesis, unless the postulated entities which are supposed to launch the microorganisms are better defined and constrained.

References

1. M. Hart, Quarterly J. Astonomical Soc. **16**, 128 (1975) reprinted in *Extraterrestrials, Where are They?* ed. by B. Zuckerman, M. Hart, 2nd edn. (Cambridge University Press, Cambridge, 1995), p. 1–8
2. http://voyager.jpl.nasa.gov/mission/fastfacts.html
3. e.g. C. Singer, *Extraterrestrials, Where are They?* ed. by B. Zuckerman, M. Hart, 2nd edn. (Cambridge University Press, Cambridge, 1995), pp. 70–85
4. For recent work by NASA to develop faster means of space transportation see for example American Institute of Physics conference proceedings, p. 813 (2006)
5. http://www.heavens-above.com/solar-escape.asp
6. E. Jones, Icarus **46**, 328 (1981) and in *Extraterrestrials, Where are They?*
7. I. Newman W, C. Sagan, Icarus **46**, 293 (1981)
8. P.-G. DeGennes, *Scaling Concepts in Polymer Physics*, (cornell University Pres, Ithaca, 1979), pp. 39–42. I am using the approximate expression $R = l(t/\tau_{hop})^{3/5}$ which follows from equation (I.23) here
9. D. Bertacchi, F. Zucca, J. Appl. Prob. **46**, 463 (2009). This paper is quite technically mathematical but contains a clear definition at the beginning
10. M.J. Fogg, *Terraforming: Engineering Planetary Envionments, Societ of Automotive Engineers.* (Warrendale, MI, 1995); C.P. McCay, Sci. Am. Presents **10**, pp. 50–52 (1999); J. M. Graham AIP Conference Proceedings, vol. 654, p. 1284 (2003)
11. This argument is not original but I have been unable to trace where I found the form of it given here. See J.R. Gott III, Nature **363**, p. 315 (1993); J. Leslie, Mind **1992** (101), 543 (1992) and references therein
12. N. Bostrom, J. Evol. Technol. **9**, March (2002). This article does not estimate probabilities per unit time
13. http://neo.jpl.nasa.gov/risk/
14. Images courtesy of Lance Benner, JPL-NASA
15. S. Arrenhius, *Worlds in the Making* (Harper and Row, NY, 1908)
16. F.H.C. Crick, L.E. Orgel, Icarus **19**, 341 (1973)
17. H.P. McSween Jr, Meteoritics **29**, 757 (1994)
18. D.D. Bogard, P. Johnson, Science **221**, 651 (1983)
19. R.H. Becker, R.O. Pepin, Earth Plan. Sci. Lett. **69**, 225 (1983); K. Marti et al., Science **267**, 1981 (1995)
20. D.S. McKay, E.K. Gibson Jr., K.L. Thomas-Keptra, H. Val, C.S. Romanek, S. J. Clemett, X.D.F. Chilier, C. R. Maaching, R.N. Zare, Science **273**, 924 (1996)
21. B.L. Kirkland, F.L. Lynch, M.A. Rahnis, R.L. Folk, I.J. Molineux, R.J.C. McLean, Geology **27**, 347 (1999)

22. See lecture by David McKay, http://www.youtube.com/watch? v = _x60v3E − yZ8
23. F.A. Rainey, K. Ray, M. Ferreira, B.Z. Gatz, M.F. Nobre, D. Bagaley, B.A. Rash, M.J. Park, A.M. Earl, N.C. Shank, A.M. Small, M.C. Henk, J.R. Battista, P. Kampfer, M.S. da Costa, Appl. Environ. Microbiol. **71**, 5225 (1979)
24. R.J.M. Fry, J.T. Lett, Nature **335**, pp. 335–365 (1988)

Chapter 7
Electromagnetic (SETI) Searches

We have argued that UFO reports do not constitute a body of scientific fact useful for evaluating the number of extraterrestrial civilizations in the galaxy. However we found that the absence of evidence of colonization of earth by extraterrestrials could be used to put some bounds on the possible types of civilizations that might exist and, in particular was shown to indicate that civilizations with the desire and ability to migrate quickly between stars and living as much as a few million years probably do not exist. We will discuss what can be learned from the negative results of spacecraft sent to other planets of our solar system in the Chap. 8. Here we focus on another major source of direct observational information relevant to the subject, namely the data collected by astronomers. We will show in this case as well that, though such data has yielded negative results with regard to observation of civilizations, this null result can also be used to put some limits on the kinds and numbers of civilizations that could exist.

To date, astronomers do not travel to other stars, though they now frequently launch instruments which take data from outside the earth's atmosphere, so they can detect various kinds of radiation. These include the entire electromagnetic spectrum as well as massive particles such as protons and electrons and nearly massless neutrinos. There have been some proposals to use studies of neutrinos incident on the earth to search for extraterrestrial civilization [1]. However, the radiation most commonly utilized for searching for extraterrestrial intelligence is the electromagnetic spectrum. It consists of waves, whose existence was first predicted by Maxwell in the mid nineteenth century, containing oscillating electric and magnetic fields. Historically the waves were first artificially generated with wavelengths of the order of meters by Hertz, guided by Maxwell's theory. About the same time, evidence was found to show that visible light was the same kind of wave, but with a shorter wave length (approximately 10^{-6} m). In the ensuing century and a half a vast technology has been developed, primarily for communication, using electromagnetic waves from short wave length x-rays ($10^{-10 \text{ to } -9}$ m), through ultraviolet (10^{-8} m), visible ($10^{-7 \text{ to } -6}$ m), infrared ($10^{-5 \text{ to } -3}$ m) and microwave ($10^{-2 \text{ to } -1}$ m) waves to radio waves(longer wave lengths to many kilometers) as illustrated in Fig. 7.1.

J. W. Halley, *How Likely is Extraterrestrial Life?*, SpringerBriefs in Astronomy, DOI: 10.1007/978-3-642-22754-7_7, © The Author(s) 2012

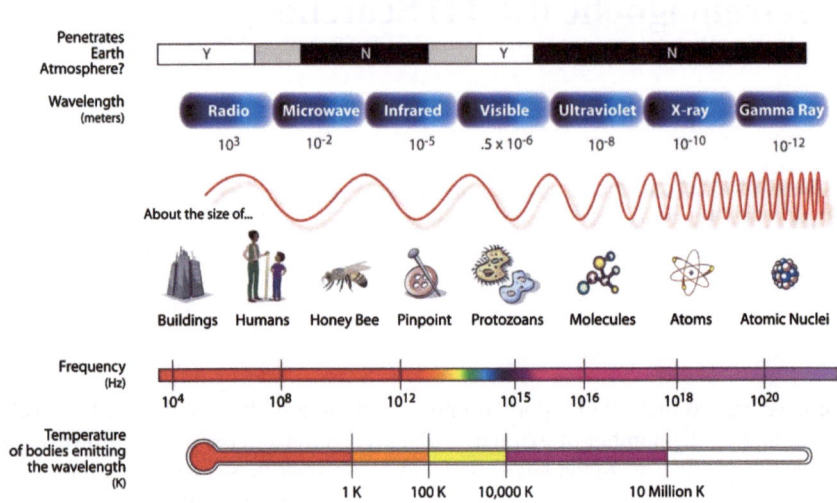

Fig. 7.1 The electromagnetic spectrum

Astronomers have concurrently built instruments to detect the corresponding radiation emitted from the skies at all these wavelengths. This process began, of course, in ancient times with visible observations of the skies, before the electromagnetic nature of visible light was known.

The advantage of using electromagnetic radiation is that, at least at certain wavelengths, it travels with minimal attentuation through the interstellar medium and the earth's atmosphere. The disadvantages of optical wavelengths associated with attenuation due to clouds is well known, but optical studies intended to investigate the question have been reported for at least 150 years, focussing in the nineteenth century on optical searches of the planets for signs of life. The twentieth century electromagnetic search for extraterrestrial intelligence (SETI) focussed mainly on the radio portion of the electromagnetic spectrum and began [2] in the mid 1950s. Drake's equation, discussed in the first part of the book, was conceived to provide a context for estimating the likelihood of success in these searches.

The community of scientists involved in the twentieth century electromagnetic searches usually assumed, in terms of the discussion in previous chapters, that $f_{life} \approx 1$ in the Drake equation, giving estimates for the number of civilizations in the galaxy of the order of a few million (Arguments for much smaller values of f_{life} appear in Chap. 4). They also argued against the conclusions of Chap. 6, contending that interstellar travel was too difficult to make colonization likely by these civilizations. For these reasons the community has been much more optimistic about the possibilities of success of the electromagnetic searches than would be implied by the discussion in Chaps. 4 and 6. This optimism led to quite detailed speculation, for

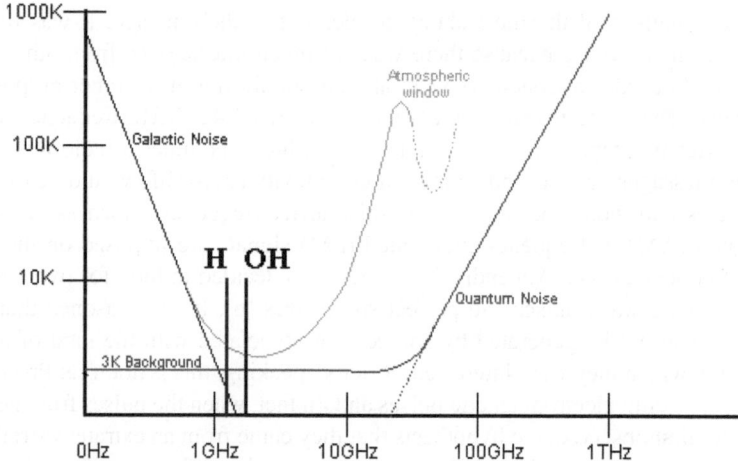

Fig. 7.2 The intensity of natural background electromagnetic 'noise' in the galaxy, showing the 'water hole' Image from Ref. [5]

example, about the types of advanced civilizations that might exist [3] (There appear to be some inconsistencies in postulating that extremely technologically advanced civilizations such as those imagined in [3] exist and are at the same time incapable of colonization which we would be likely to observe as discussed in Chap. 6). One of these speculations, due to Freeman Dyson, was that sufficiently advanced civilizations would completely enclose their parent stars in reflecting material to maximize stellar energy collection. Such 'Dyson spheres' could be conjectured to radiate in the infrared (somewhat like the surface of a planet as discussed in Chap. 3). At least one search for such objects in data from an orbiting space based infrared telescope has been reported [4], with negative results.

Radio waves in the meter wave length range are less attenuated by the interstellar medium and the earth's atmosphere than those from other parts of the spectrum, and the advent of radio astronomy in the mid twentieth century therefore provided new opportunities. One can understand how radio astronomers have chosen the range of frequencies and wavelengths to use in their search from Fig. 7.2, which exhibits the intensity of the known sources of electromagnetic radiation in the radio frequency regime.

The so called 'water hole' labelled in the spectrum contains several frequencies at which electromagnetic radiation is characteristically absorbed and emitted by water molecules. Particular attention has centered on the frequency of 1,420 MHz (1.420×10^9 cycles/s) corresponding to a wavelength of electromagnetic radiation of 21 cm. This radiation arises from the dynamics of the magnetic moments of the proton and the electron in the hydrogen atom (See Appendix 7.1). It is used by radio astronomers to map the presence of hydrogen in the universe. SETI researchers argued that an advanced civilization might choose it as a natural medium of communication because it would be presumed that other civilizations would be familiar with the dynamics of the most abundant element in the universe. It was furthermore convenient to use because there was not much interference from other noise sources and because astronomers were already monitoring it for other purposes in astronomy. The SETI researchers assumed that this 1,420 MHz frequency would be a 'carrier frequency', like the frequency to which one tunes a radio or TV and that any messages or other indications of complexity and/or life would be of lower frequencies and would be imposed on this carrier frequency, much as amplitude modulated (AM) or frequency modulated (FM) signals are imposed on the radio carrier frequencies (see Appendix 7.2). They also decided to look for two kinds of signals: square wave pulses and perfect sine waves [6]. It was reasoned that such signals would not be generated by sources not associated with the kind of intelligent life in which they were interested. Strictly speaking, this is not true: Precessing neutron stars emit electromagnetic pulses and, in fact, when the pulses from neutron stars were first observed, the hypothesis that they came from an extraterrestrial civilization was briefly considered [7] until a more mundane (though still quite exotic) explanation was found [8]. This basic search strategy has not changed qualitatively during the 50 years of SETI searches which have occurred, though the sensitivity of the detectors and the 'bandwidth' of frequencies around 1,420 MHz which are searched have significantly increased.

For efficient detection, the telescopes for radio waves need to have reflecting 'lenses' of size comparable to or larger than the wavelengths of interest. Consequently the instruments are dramatically large. In Fig. 7.2 we show a picture of the radio telescope at Arecibo, Puerto Rico, which has been used for many Search for Extraterrestrial Intelligence studies . The 'Allen telescope', of which a picture is shown in Fig. 7.4, is an array of dish antennae which are electronically linked to detect coherent signals over a large area. The Allen telescope has been devoted almost exclusively to SETI searches and was built with privately donated funds. Unfortunately, the Allen telescope recently (April 2011) suspended observations because its operators had inadequate funds to continue. SETI efforts have had fluctuating financial support throughout their history in the United States. Most of the effort has been supported by private entities with a brief interlude of public support in 1991–1992. Some of this history is reviewed in [9, 10].

The number of existing searches for electromagnetic evidence of the existence of extraterrestrial civilizations is now quite large. Some data on recent searches is provided in Appendix 7.3. (The data were assembled from Ref. [11] but unfortunately this website no longer posts such a convenient tabular summary). In the Appendix 7.3, the 'band width' is the frequency or wavelength range of the signals which the antenna or telescope can detect. As discussed, the frequency ranges were chosen by the observers because the level of natural electromagnetic background noise in the galaxy is low at those frequencies.

Quite considerable resources have been expended to scan the sky for objects emitting pulses at frequencies near 1,420 MHz. Because of complications associated with the fact that emissions from an orbiting planet at this frequency would be Doppler shifted, various computations have been performed on the data to discover if any of them can be interpreted as pulses from an orbiting planet. The needed computational

Fig. 7.3 Top: The 305 m radio telescope in Arecibo, Puerto Rico [13]. Bottom: The Allen telescope array in Hat Creek, California [14]

resources are immense because a huge amount of data has been collected, particularly in 'parasitic' searches which collect radio telescope data while the telescope is being used for other astronomical work. An innovative computational program called SETI@home has been used to link more than a million personal computers to carry out this work in one of the projects. In spring 2011, SETI@home resources are analyzing data collected by a radio telescope at Green Banks, Virginia from 86 stars identified by the Kepler satellite as being likely to be orbited by earth-like planets [12]. The earlier seti@home searches covered about 1/3 of the sky with a sensitivity of around 10^{-25} W/m^2 (the minimum strength of signal which the antenna could detect). No signals of the sort sought (pulses or sine waves) were identified which could not be attributed either to natural processes or to human activities. The sensitivity should be adequate to detect a signal from a sunlike star anywhere in the galaxy, assuming that the civilization devotes a small fraction of the power of the star to sending the signal (This does not take account of the fact that signal from the galactic center would be blocked) (Fig. 7.3).

7.1 Effect of the Finite Time for Propagation of Electromagnetic Signals

The time for propagation of light or radio signals from remote parts of the galaxy or other parts of the universe can be extremely long (up to the age of the universe itself). Thus we need to be concerned with how to take account of the propagation time in analysing the results of SETI searches. Here I will only consider signals from our own galaxy in the discussion, though it could be easily extended to searches of the entire observable universe. Suppose that the lifetime of a civilization is, on average, τ_d. For our purposes here we think of this as the 'electromagnetic' lifetime during which the civilizations puts out SETI detectable signals. If the size of the galaxy is R, then the time for the signals to leave the galaxy is of order R/c. The distribution of electromagnetic signals from a civilization now depends on whether $R/c > \tau_d$ or $R/c < \tau_d$. ($R/c \approx 10^4$ years) If $R/c < \tau_d$ then there will be a period during which each civilization fills the galaxy with its signal, while this will never happen if $R/c > \tau_d$. In the latter case, the signal will fill a sphere for a time τ_d and then it will fill a shell in the galaxy of approximate volume $\tau_d c R^2$ for an additional time of order R/c. However this does not alter our estimate of the probability of detection (which we take to be N_{civ}/N_{gal} where N_{civ} is the average steady state value of the number of civilizations alive at any moment): Consider a given star, which SETI observes. To see an electromagnetic signal from a civilization on it, it is necessary that it was 'alive' a time r/c ago where r is the distance to the star. If we assume that each star is only 'alive' once and that the time it is alive is evenly distributed over the lifetime τ_u of the universe, then the probability that it was alive a time r/c ago is τ_d/τ_u times the probability that it was ever alive. The probability that it was ever alive, given a birthrate per star of $1/\tau_{birth}$ is τ_u/τ_{birth} as long as $\tau_u < \tau_{birth}$ Assuming $\tau_u < \tau_{birth}$ for now (see below) we obtain that the probability that the star being observed is seen to be emitting a signal is $(\tau_d/\tau_u) \times (\tau_u/\tau_{birth}) = \tau_d/\tau_{birth}$. But in a steady state the birth rate of civilizations equals the death rate so that

$$N_{civ}/\tau_d = N_{gal}/\tau_{birth}$$

giving

$$(\tau_d/\tau_{birth}) = N_{civ}/N_{gal}$$

(this form of the Drake equation was also mentioned in the last chapter and in Appendix 1.1) so the probability of observing a civilization on a star remains N_{civ}/N_{gal} even if the lifetime of civilizations is less than R/c, as long as $\tau_u < \tau_{birth}$. Whether the latter condition holds is not known. If $\tau_u > \tau_{birth}$ then the probability that the star was ever 'alive' becomes (of order) one and the probability that the observed star is seen to be emitting a signal becomes (τ_d/τ_u) which in that case would be smaller than N_{civ}/N_{gal}. For 'optimistic' scenarios in which N_{civ} is of order 10^6 and τ_d is of order 10^6 yr, τ_{birth} is of order $10^6 \times 10^5 = 10^{11}$ yr which is quite close to (but larger than) τ_u. However, our discussion of prebiotic evolution in Chap. 4 suggested much larger values of τ_{birth}. In either case, it appears that $\tau_u < \tau_{birth}$ is the more likely case and then the probability of SETI success per star is N_{civ}/N_{gal} even when civilization lifetimes are less than R/c. Notice that our own electromagnetic lifetime, so far, is 10^2 yr. We discussed the probability that this will become much longer in the last chapter.

7.2 Bounds on N_{civ} From the SETI@Home Search

Now suppose that we are looking electromagnetically for civilizations that emit a particular kind of electromagnetic signal which we can detect from anywhere in the galaxy. It is quite universally agreed that the average number of civilizations N_{civ} in a galaxy is much less than the average number of stars in a galaxy N_{gal} so we will make this assumption in the analysis. If we have examined a fraction f of the N_{gal} stars in the galaxy and there are N_{civ} civilizations of this type in the galaxy we can write a formula for the probability that we have failed to observe one of them:

$$P = (1 - N_{civ}/N_{gal})^{f N_{gal}}$$

To think about this, it is convenient to let $x = N_{civ}/N_{gal}$ and write it as

$$P = (1 - x)^{(1/x)f N_{civ}}$$

x, is the probability of finding a civilization on a randomly selected star. It is generally agreed to be $<< 1$. We can write the last expression as

$$P = \left[(1 - x)^{1/x}\right]^{f N_{civ}}$$

The quantity $(1 - x)^{1/x}$ when x gets small takes a definite value, usually called $e^{-1} = 1/2.71828\ldots$ so we have

$$P = e^{-f N_{civ}}$$

For the SETI@home search, f is around 1/3. If we put that value into the formula for the probability P of failure to observe a civilization with the characteristics sought we can plot the result as a function of the number N_{civ} of civilizations. The result is shown in Fig. 7.4.

It is clear that if the number of civilizations of the type sought in the galaxy were more than about ten (and they were evenly distributed) then the probability that they have been missed by a SETI search of this type is extremely small.

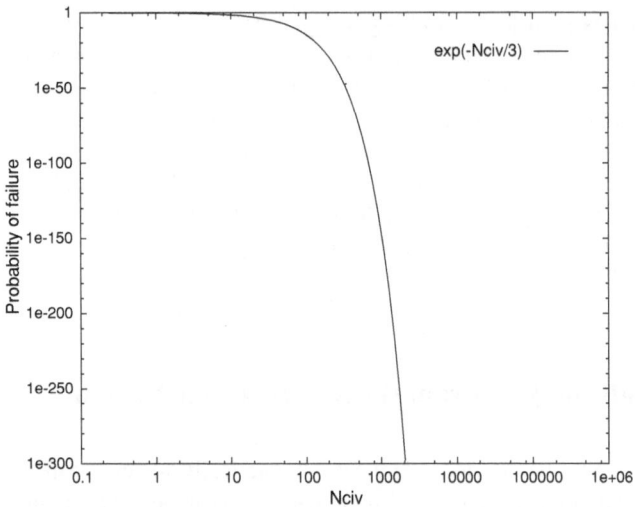

Fig. 7.4 Probability that all sky Arecibo searches have failed to find N_{civ} civilizations emitting signals of the type and intensity for which they were sensitive, as estimated in the text

This estimate of the likelihood of failure to observe an electromagnetic signal is less significant than it would be if it were not for the fact that the sky survey SETI@home search has confined attention to an extremely narrow band width and to quite limited kinds of analogue signals [6] imposed on carrier frequencies in the near neighborhood of 1,420 MHz.

7.3 Possible Alternative Signal Processing Strategies

Human civilization (the only one about which we know anything) is not regularly transmitting pulsed or sine wave signals, either directed or isotropically, into the galaxy on a carrier frequency of 1,420 MHz or anything near it, though brief signals have been sent to around 20 stars in various 'active SETI' programs [15]. In the human case this is firstly because human societies have other more pressing priorities and secondarily because, on the chance that extraterrestrial civilizations are present, sending signals might be imprudent from a security point of view [16]. This might suggest possible flaws in the reasoning which led to searches for sine wave and pulsed signals at around 1,420 MHz, and it might therefore be of interest to consider some alternative strategies for looking for signs of life in the radio data which is being collected. Here I briefly discuss four possibilities: (1) signals with characteristics like the electromagnetic emissions of humans from earth, (2) Anomalous power laws in the radio emissions at low frequencies, (3) signals with statistical characteristics like those of human speech, or (4) signals with signatures characteristic of complex phenomena more generally. Each of these possibilities is briefly discussed next.

1. The most straightforward approach is to ask what kinds of anthropogenic electromagnetic signals the earth is currently broadcasting. Though this is certainly a somewhat parochial approach, it has the merit of using data from a real biosphere and is arguably more likely to yield a useful strategy than attempts to guess what an undefined and completely unknown 'advanced' biosphere might do. Some data is available in the literature concerning the spectrum of anthropogenic electromagnetic emissions from the earth, though the published data is less extensive than one might expect. A review of the humanly generated radio frequency radiation from earth as of 1978 was discussed in the SETI context in Ref. [17]. An attempt was made [18], in 1990 to detect biogenic signals from earth by the Galileo space mission (discussed in the Chap. 8). A few narrow band impulsive radio emissions were detected. In 1996, more detailed anthropogenic data from the earth in the frequency range 0 to 14 MHz from the WAVES/Wind satellite was reported [19]. The authors noted that the intensity distribution at fixed frequency observed from the human generated sources was qualitatively different from that observed from the nonbiogenic auroral kilometric radiation coming from earth at nearby frequencies as shown in Fig. 7.5.

2. Statistical characteristics of humanly produced *sound* were determined experimentally decades ago [20] Over quite a wide range of low frequencies, the spectrum of human sound amplitude rises monotonically with decreasing frequency, approximately as a power law. (Power $\propto 1/f^{\alpha}$, with α of order 1). Data from Ref. [20] taken from radio broadcasts, is shown in Fig. 7.6. It has been suggested that such an approximately $1/f$ emission spectrum might be characteristic of the kind of complexity characteristic of advanced forms of life. In fact, $1/f$ noise does appear in other biological contexts [21, 22]. But using this idea to formulate a useful strategy for a SETI search is complicated by the fact that $1/f$ noise is produced by many natural systems [23, 24] that are not lifelike, so discovery of such an emission spectrum coming from a star will not unambiguously identify the star as harboring any lifelike entities. However the presence of such an emission feature could possibly serve as a necessary (though not sufficient) condition for regarding a star as possibly harboring complex life.

3. Statistical characteristics of written human language have also been analysed using several different mathematical techniques. They all suggest very long range 'correlations', in which the content of the message at one instant of time depends, on the average, on the content of the message at previous times, with a declining influence of content in the more distant past. Such correlations affect the low frequency spectrum if the resulting signal is converted to a string of numbers and analysed for its spectral content. There are some subtle issues associated of how to identify the symbols of a message in human language with numbers which are discussed, for example in Ref. [25] and, briefly, in Appendix 7.4. Li, in Ref. [25] and resolving the issues in a particular way,

Fig. 7.5 Distribution of intensity of humanly generated radio signal a 7.25 MHz (left) compared with (right) the intensity distribution from the auroral kilometric radiation (AKR) from earth at 0.25 MHz. From Ref. [19]

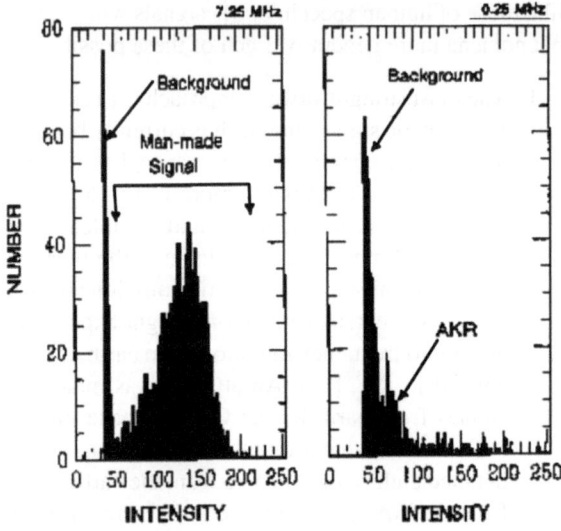

found correlations decreasing as approximately 1/(time in the past) and giving a very weak low frequency dependence in the spectral function. An analysis of human language using one of the definitions of complexity briefly discussed next has been given by Grassberger [26].

4. Another possible approach to formulating hypotheses concerning the kinds of signals to seek in a SETI search would be to seek signals characteristic of some sort of quantitatively defined nonequilibrium 'complexity'. There has been a lot of work on analysis of the complexity of signals in the context of electrical engineering and biology, but none of it has, to my knowledge, been applied to data from stellar objects. Many of the proposed measures use the information content $I = \log_2 W$ where W is the total number of possible messages. If an electromagnetic signal is observed as a function of time, then, by using a binary code to represent the numbers giving the amplitude of the signal as a function of time, it can be translated into a series of 0 and 1s. If the resulting signal is N units long and if every possible sequence of N 0 and 1s is allowed as a message, then the information content of any message is $I = N \log_2 2 = N$. If some sequences of 0 and 1s are not allowed, which is the usual case in real messages, then the number of possible messages is smaller and the information content of any given message is smaller.

To understand that the number of allowed messages is less than 2^N in any sensible message, consider a person typing a message of N' letters of English. Allowing spaces but no capitals or punctuation, the number of possible sequences is $27^{N'}$ but most of these would be gibberish and would not be counted (or typed) in English so the total number of allowed messages is less than $27^{N'}$. Another way to describe this situation is to say that ordinary human languages do NOT maximize information

Fig. 7.6 Log-log graph of
the measured sound
spectrum from various radio
stations [20]

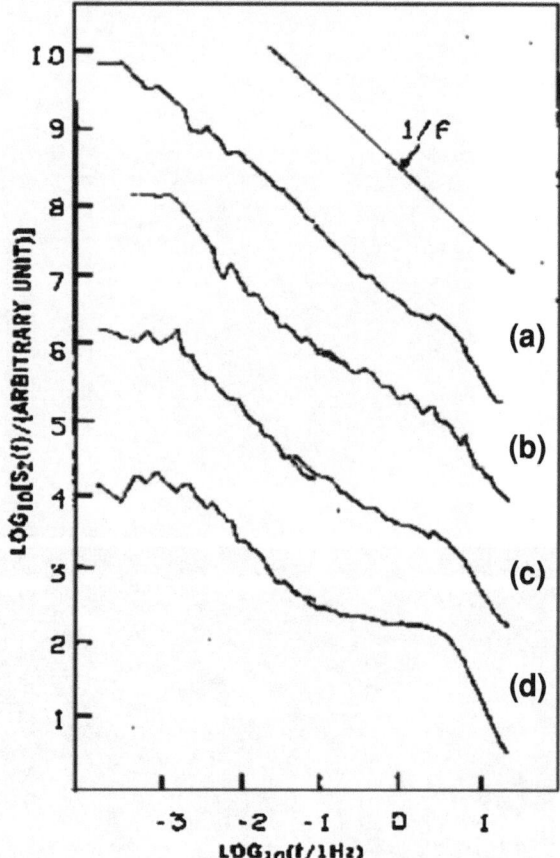

content. In fact it appears that a message which DID maximize information in its
messages would be impossible to recognise without knowing the code in advance
(This point was made in the context of SETI by Lachmann et al. [27] where the
constraint that the message makes maximally efficient use of the available energy is
added. But the conclusion is essentially the same).

Without the energy constraint taken into account in Ref. [27] the frequency spectrum of a message which maximized information content per unit time would show
the same amplitude at every frequency. That is it would be completely 'white' noise.
This is consistent with another of idea how to define complexity: one proposes to
define complexity as 'computational depth' by which one means the length of the
shortest computer program required to generate the message. In the case of apparently random messages of maximal information content, the only code which will
generate them is basically a list of all the possible messages, which is of maximal
length.

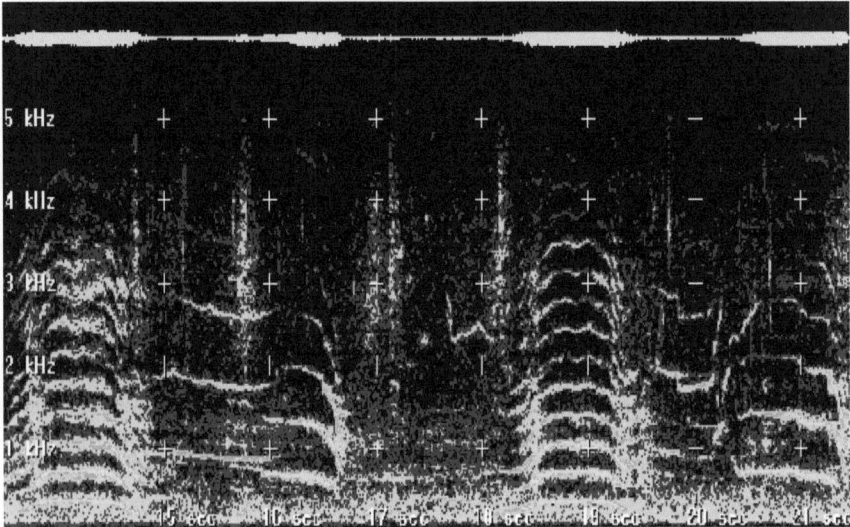

Fig. 7.7 *Top* A humpback whale [29]. *Bottom* Sound spectrum from the song of the humpback whale measured at 1 s intervals [30]

If information transmission per unit time in human messages is not being opti-
mised, what is? This does not seem to be known precisely, but it seems reasonable
to conjecture that it is some kind of communicability. By inserting the repetitions
and correlations which are required by the rules of grammar into a message, it is
made easier for the receiver to interpret, though the information content goes down.
Given some message, it is possible in several different ways to count the number of

repetitions to keep track of how frequent and close together the repetitions are. Some of the ways of doing it are reviewed in Appendix 7.4. The methods (algorithms) can be applied to messages for which the code is not known. This has been done for English texts of various kinds [25] as well as for the songs of humpback whales [28]. Basically what is found is that in human messages the information content is smaller than the alphabet permits and the messages are correspondingly 'correlated', which means that if a letter appears, than only some of the letters in the alphabet are likely to appear next (spelling rules) and if a word appears then some words are more likely to appear next than others (rules of grammar) and so forth out to very long correlations, such that the end of a novel, for example, usually is not randomly related to the beginning of the novel. There are indications (Fig. 7.7) that similar structures appear in the songs of the humpback whale [28] (though we have no idea what, if anything, they might mean). These quantitative algorithms for evaluating the complexity of signals could be used to evaluate signals from stars containing earthlike planets to see if they suggest complexity different from that observed in most natural processes. This does not appear to have been done.

7.4 Conclusions From the Survey of Results From Electromagnetic Searches

In summary, the negative results of SETI sky survey searches strongly suggest that less than 100 (and probably less than 10) civilizations emitting signals of the type sought exist in the galaxy. However in view of the very limited kind of signal for which these searches looked, it would seem advisable to search for a wider range of signal types. This is being done by widening the band width of the searches, but the basic kinds of signals sought are still very limited. A variety of other algorithms might be used to analyse the data gathered in SETI searches. We have mentioned the possibility of seeking stars with $1/f^{\alpha}$ electromagnetic emission spectra or exhibiting correlations like those observed in human communications.

References

1. J.G Learned, S. Pakvasa, W.A. Simmons, X. Tata, Q.J. Roy, Astron. Soc. **35**, 321–329 (1994)
2. G. Cocconi, P. Morrison, Nature **184**, 844–846 (1959)
3. N.S. Kardashev, Sov. Astron. **8**, 217–221 (1964)
4. D. Carrigan, Int. Astronaut. Cong. IAC-04-IAA-1.1.1.06
5. SETI League image, http://www.setileague.org, used by permission
6. A rather old review of signal processing methods appeared in D. Kent Cullers, I.R. Linscott, B. Oliver, Commun. ACM **28**, 1151–1163 (1985)
7. http://www.csupomona.edu/nova/scientists/articles/burn.html
8. A. Hewish, S.J. Bell, J.D.H Pilkington, P.F Scott, R.A Collins, Nature **217**, 709713 (1967)
9. S.J. Garber, J. Br. Interplanet. Soc. **52**, 3–12 (1999)

10. http://www.csupomona.edu/nova/scientists/articles/burn.html
11. http://www.seti.org
12. D. Perlman, San Francisco Chronicle 17 May (2011)
13. Courtesy of the NAIC – Arecibo Observatory, a facility of the NSF
14. Allen Telescope Array copyright (2010) SETI Institute, Courtesy SETI Institute
15. http://en.wikipedia.org/wiki/ActiveSETI
16. http://iaaseti.org/smiscale.htm
17. W.T. Sullivan III, S. Brown, C. Wetherill, Science **199**, 377 (1978). I am grateful to Jill Tarter for pointing out this paper
18. C. Sagan, W.R. Thompson, D. Gurnett, C. Hord, Nature **365**, 715 (1993)
19. M.L. Kaiser, M.D. Desch, J.L. Bougeret, R. Manning, C.A. Meetre, Geophys. Res. Lett. **23**, 1287 (1996)
20. R. Voss, J. Clarke, J. Acoust. Soc. Am. **63**, 258 (1978)
21. G.J. Van Orden, J.G. Holden, M.T. Turvey, Exper. Psychol.: Gen. **132**, 331–350 (2005)
22. I. Lundstrm, D. McQueen, J. Theoret. Biol. **45**, 405–409 (1974)
23. M.B. Weissman, Rev. Mod. Phys. **60**, 537–571 (1988)
24. W.H. Press, Comments Astrophys. **7**, 103–119 (1978)
25. W. Li, Complex Syst. **5**, 381 (1991)
26. P. Grassberger, IEEE Trans. Inf. Theor. **35**, 669 (1989)
27. M. Lachmann, M.E.J. Newman, C. Moore, Am. J. Phys. **72**, 1290 (2004)
28. R. Suzuki, J.R. Buck, P. Tyack, J. Acoust. Soc. Am. **119**, 1849 (2006)
29. Photograph by Brandon Cole. Used by permission
30. Figure from reference [28], copyright American Institute of Physics, used by permission

Chapter 8
Direct Searches for Primitive Forms of Life

While electromagnetic searches have the advantage that they can be conducted without actually sending a material object or device to a suspected host of extraterrestrial life, there are also limitations to this method of exploration. The biosphere or organisms may not produce observable radiation and some kinds of experiments, such as chemical analysis of soils, which are routine in studies of terrestrial life, are not possible. The NASA astrobiology program and related programs in other countries complements SETI searches by sending robotic probes to the planets and moons of our solar system. The limitation to our solar system is imposed by practical limitations as we have discussed in Chap. 6. Sending material probes to other stars would take about 10^4 years or more using present technology. (NASA does contemplate long term development of an interstellar probe). Here we briefly discuss results from such studies of Mars, Europa, a moon of Jupiter and Titan, a moon of Saturn. These are the objects of greatest current interest in the astrobiological efforts of NASA (Fig. 8.1).

8.1 Mars

By the criteria discussed in Chap. 2, the other planets in our solar system do not appear to be very habitable. In particular the surface temperatures of Mercury and Venus are too high and those of Mars and the outer planets are too low to permit the presence of liquid water on their surfaces. Before any space probes had visited these planets in the 1960s, the most promising case was thought to be Mars. Planning for Mars exploration began in the 1950s with the search for life on its surface as a significant scientific objective. The first close up images of the Martian surface were provided [1] in 1965 by the Mariner 4 mission (Fig. 8.7).

> It took seven and a half months to travel the 525 million km to Earth's neighbour. The 260 kg spacecraft began its brief encounter with the planet on 14 July 1965. Among other measurements, the vidicon television system during a 25 min sequence took 21 full pictures and a fraction ... of the Martian surface at distances of 10,000 to 17,000 km. After being stored overnight on a tape recorder, the images were transmitted to Earth the next day [2].

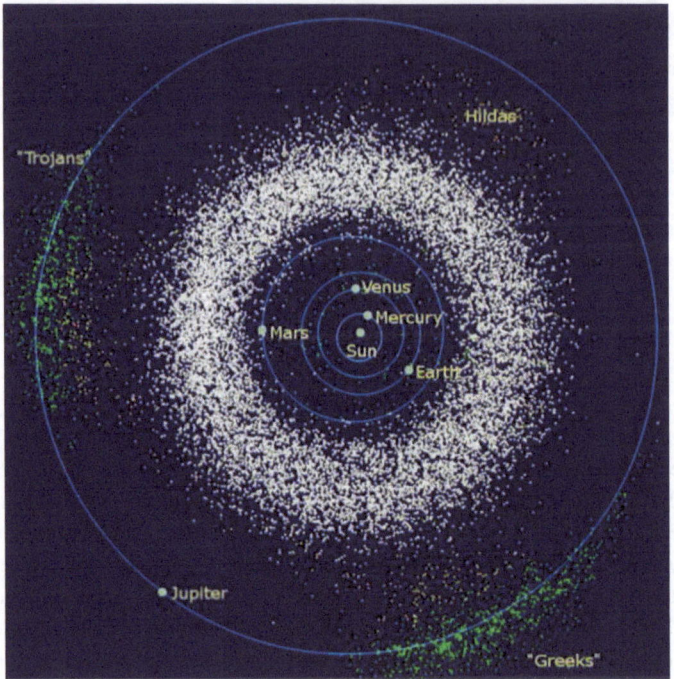

Fig. 8.1 The *inner* solar system

Scientists involved in the Mariner mission and particularly the "exobiologists" who were planning the search for life on Mars were surprised and reportedly disappointed by many of the results [2]:

From the pictures (taken by Mariner 4), the TV team thought some fundamental inferences could be drawn:

1. In terms of its evolutionary history, Mars is more Moon-like than Earth-like. Nonetheless, because it has an atmosphere, Mars may shed much light on early phases of Earth's history.
2. Reasoning by analogy with the Moon, much of the heavily cratered surface of Mars must be very ancient-perhaps two to five billion years old.
3. The remarkable state of preservation of such an ancient surface leads us to the inference that no atmosphere significantly denser than the present very thin one had characterized the planet since that surface was formed. Similarly, it is difficult to believe that free water in quantities sufficient to form streams or to fill oceans could have existed anywhere on Mars since that time. The presence of such amounts of water (and consequent atmosphere) would have caused severe erosion over the entire surface.
4. The principal topographic features of Mars photographed by Mariner have not been produced by stress and deformation originating within the planet, in distinction to the case of the Earth. Earth is internally dynamic giving rise to mountains, continents, and other such features, while evidently Mars has long been inactive. The lack of internal activity is also consistent with the absence of a significant magnetic field on Mars as was determined by the Mariner magnetometer experiment.

Fig. 8.2 The Mariner 4 spacecraft. Images from NASA [3]

Fig. 8.3 First Mariner 4 pictures of the surface of Mars (1965). Images from NASA [3]

5. As we had anticipated, Mariner photos neither demonstrate nor preclude the possible existence of life on Mars. The search for a fossil record does appear less promising if Martian oceans never existed. On the other hand. if the Martian surface is truly in its primitive form, the surface may prove to be the best-perhaps the only-place in the solar system still preserving clues to original organic development, traces of which have long since disappeared from Earth.

Image of the Mariner 4 spacecraft appear in Fig. 8.2 and photographs which it made of the Martian surface are shown in Fig. 8.3. The Mars surface was found to be colder ($-100 \pm 20°$C) than earlier estimates ($30 \pm 50°$C estimated in 1964) and and

Fig. 8.4 Viking landing sites [3] on Mars (1976)

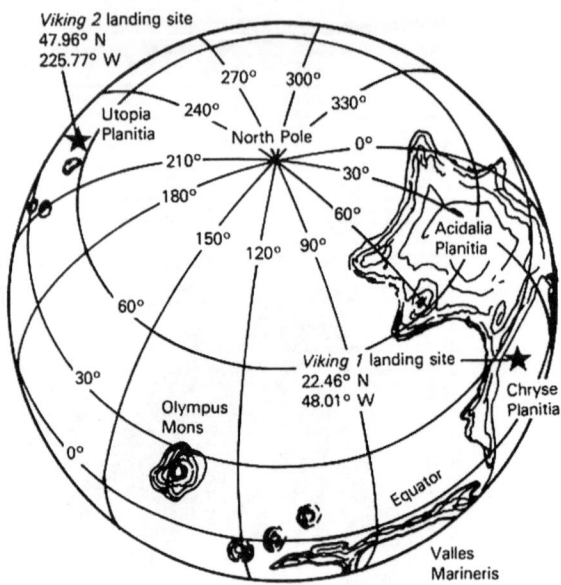

its atmosphere was thinner (pressure 5 millibars versus earlier estimates of 20 ± 10 millibars).

Mariner 4 was followed by several more Mariner missions [6, 7, 9 (the first orbiter)] which gathered more data on Mars including a complete map of the surface by Mariner 9, but none of which landed a robotic probe on its surface. The first successful human probes to land on the surface of Mars were provided by the Viking mission which left earth in 1975 and put two functioning landers on the Martian surface in summer of 1976. (Several Soviet attempts preceded them and one provided brief telemetry but little new information about the surface before failing). The scientists engaged in the design of the Viking lander experiments included at least one Nobel prize winner (Joshua Lederberg), Carl Sagan of Cornell, who became a major public figure commenting on the search for extraterrestrial life, and National Academy of Sciences member Al Nier of Minnesota, who had first isolated the fissioning isotope U^{235} during the Manhattan project during World War II. The landers were explicitly and primarily designed to look for signs of life. The probes were very carefully purged of terrestrial microbes so that, if it detected any indications of microorganisms in the Martian soil, one could be sure that they were of Martian origin [3]. The landing sites are shown in Fig. 8.4.

Each Viking Lander carried an integrated biology instrument, which contained three experiments designed to detect the metabolic activity of microorganisms should they be present in the soil sampled. First, the gas-exchange experiment would determine if changes caused by microbial metabolism occurred in the composition of the test chamber atmosphere. Second, the labeled-release experiment, also known as Gulliver, would determine if decomposed organic compounds were produced by microbes when a nutrient was added. Third, the pyrolytic-release experiment would detect, from gases in the chamber, any synthesis of

Fig. 8.5 Viking picture [3]
of the Martian surface

organic matter in the Martian soil. A change could be the result of either photosynthetic or
nonphotosynthetic processes.

A picture of the Martian surface taken by a viking lander appears in Fig. 8.5. The
integrated biology instrument is shown in Fig. 8.6. This last instrument gave some
preliminary indications of unusual chemistry, but analysis showed, to most scientists'
satisfaction, that it was entirely due to nonbiological processes.

However a more telling result came from the gas chromatograph mass spectrometers which
were aboard the Viking landers. These instruments could detect organic molecules and were
regarded as much more definitive tests of the possibility of carbon based life on Mars either
in the present or at some time in the past [3].

On 12 August (1976), the GCMS experiment was run again with the first sample being heated
to a maximum temperature of 500°C. Biemann reported that this analysis 'to our surprise,
evolved a large amount of water. Indeed so much that it gives us trouble in analyzing the
data'. Still, the critical point of this analysis was that there were probably no organics. If
the reactions observed in the biology instrument were the consequence of life, then it was
expected that the GCMS would detect organic compounds in the same soil. Neither this
analysis nor the subsequent one at the Viking 1 site, nor those carried out at the Viking 2
landing area, produced traces of organic compounds at the detection limits (a few parts per
billion) of the GCMS.

Failure of the gas chromatograph-mass spectrometer to detect organic compounds was devastating for those who believed that life on Mars was possible. For Jerry Soffen (scientist), the
GCMS results were 'a real wipe out.' Once he assimilated the fact that the GCMS had found
no organic materials, he walked away from where the data were being analyzed saying to
himself, 'That's the ball game. No organics on Mars, no life on Mars.' But Soffen confessed
that it took him some time to believe the results were conclusive.

Fig. 8.6 The Viking *soil* sampler [3]

Thirty-two years later, most scientists believe that there is almost certainly no life on Mars (though some scientists continue to believe that the results of the labelled release experiment in the Viking missions might have indicated biological activity [4].) Very active exploration of the surface of Mars using many robotic probes has continued through the first decade of the twenty-first century. Varying estimates of the amount of water on Mars in the past and present have resulted, but there has been no indication of biological activity. It is instructive to note that scientists of the highest reputation took the possibility of Martian biology with utmost seriousness and advised the US government to invest very large sums to test the hypothesis. This is not to imply a negative assessment of their skill or judgement. But it illustrates quite dramatically that humans then had (and still have now) extremely limited understanding of the requirements for initiation of life on planets orbiting around stars.

8.2 Europa

More recently attention of space scientists interested in these issues has shifted to the moons of the outer planets. They are extremely cold, but there are indications that a

Table 8.1 Galilean moons of Jupiter [5]

Satellite	GM (km^3/s^2)	Mean radius (km)	Mean density (g/cm^3)	Magnitude V0 or R
Io	5959.916 ± 0.012	1821.6 ± 0.5	3.528 ± 0.006	5.02 ± 0.03
Europa	3202.739 ± 0.009	1560.8 ± 0.5	3.013 ± 0.005	5.29 ± 0.02
Ganymede	9887.834 ± 0.017	2631.2 ± 1.7	1.942 ± 0.005	4.61 ± 0.03
Callisto	7179.289 ± 0.013	2410.3 ± 1.5	1.834 ± 0.004	5.65 ± 0.10

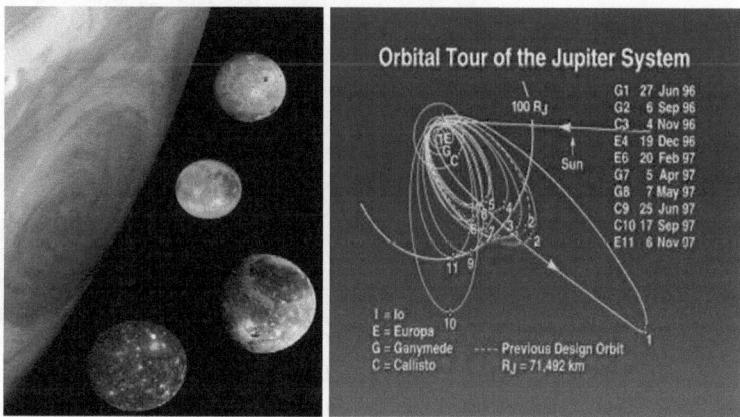

Fig. 8.7 *Left*: Images of the Galilean moons of Jupiter at correct relative sizes (but not distances) compared with the size of Jupiter itself. Images were obtained in 1996 by the Galileo spacecraft [5] *Right*: Path of the Galileo space craft

few of them may have subsurface oceans, possibly containing liquid water. Interest has centered on the moons of Jupiter and of Saturn.

More than 50 moons of Jupiter are known (depending on how small an object should be called a moon, the number is not precisely defined). However the main interest is in the four moons discovered more than 300 years ago by Galileo. Some of their properties are summarized in Table 8.1.

The relative sizes of the Galilean moons and their appearance are indicated in Fig. 8.7. New information about these moons was obtained by the Voyager missions, discussed earlier, and more recently by the Galileo spacecraft which orbited Jupiter for more than a year in 1995 and 1996. The track of Galileo and its visits to the Galilean satellites is also summarized in the Fig. 8.7.

Of the four Galilean moons, Io was found to be very volcanically active but with an atmosphere thought to be very inhospitable to life. Ganymede and Callisto were not found to contain much water. Attention focussed on Europa. Its appearance is quite dramatically different from the other three. Pictures of all four moons appear in Fig. 8.7 and a more detailed picture of Europa taken by Galileo is in Fig. 8.8.

The cracks on the surface of Europa are thought arise from repeated upwelling of a fluid, speculated to be water, from its interior. Other indications that there might be subsurface liquid water on Europa come from measurements of the magnetic field

Fig. 8.8 NASA image of
Europa [5]

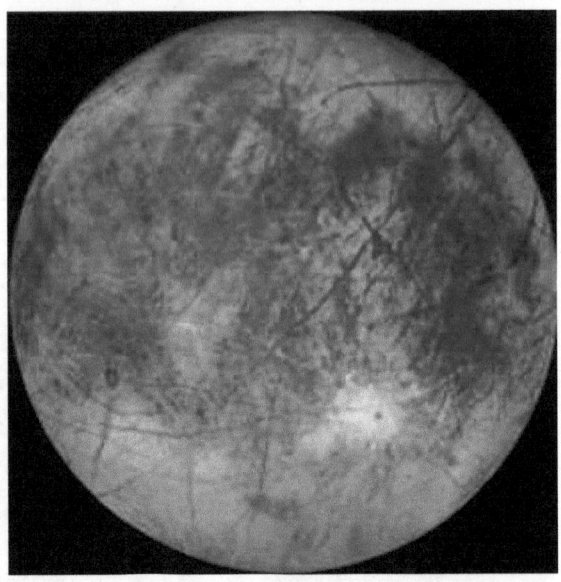

around Europa. The large magnetic field of Jupiter was found to be significantly distorted by Europa in a way consistent with the presence of a subsurface ocean of water (though other interpretations remain possible). For these reasons, further missions to Europa will include surface landers and it is hoped that it will be possible to equip such landers to penetrate the surface to ascertain whether there is indeed a subsurface ocean of liquid water. Though NASA workers are hopeful that such an ocean exists, the presence of liquid water alone does not assure the presence of life. Some further information about Europa appears in Table 8.2:

In the context of a search for life on Europa, the observed surface temperatures are relevant. They are very low. A subsurface ocean, if it exists, will have to be almost 200°C warmer. Though estimates suggest that this may be possible, it seems clear that life on Europa could only include very extreme forms of 'extremophiles'. Even if the probability of life initiation in warm aqueous ponds or ocean trenches on earth is a lot more probable than the models we have discussed suggest, to expect the harsh environment of a subsurface Europan ocean to initiate life by similar processes must be regarded as many orders of magnitude less probable. More colloquially, life on Europa looks like a very long shot. One wonders if it would be seriously considered if the astrobiology community at NASA had a smaller investment in finding life somewhere in the solar system and had not been disappointed on Mars and elsewhere.

8.3 Titan

The other moon of major interest to astrobiologists at NASA is the moon named Titan orbiting Saturn. The orbits of the larger moons of Saturn are shown in Fig. 8.9.

Table 8.2 Properties of Europa [4]

Discovery	Discovered by G. Galilei on January 7, 1610

Orbital characteristics:

  Periapsis: 664,862 km
  Apoapsis: 676,938 km
  Mean orbit radius: 670,900 km
  Eccentricity: 0.009
  Orbital period: 3.551181 d
  Average orbitalspeed: 13.740 km/s
  Inclination: 0.470° (to Jupiter's equator)

Satellite of: Jupiter

Physical characteristics:

  Mean radius: 569 km (0.245 Earths)
  Surface area: 3.09×10^7 km^2 (0.061 Earths)
  Volume: 1.593×10^{10} km^3 (0.015 Earths)
  Mass: 4.80×10^{22} kg (0.008 Earths)
  Mean density: 3.01 g/cm^3
  Equatorial surface gravity: 1.314 m/s^2 (0.134 g)
  Escape velocity: 2.025 km/s
  Rotation period: Synchronous
  Axial tilt: 0.1°
  Albedo: 0.67 ± 0.03
  Surface temp: min 50 K; mean 103 K; max 125 K
  Surface pressure: 1 Pa

Saturn has been visited by the Voyager spacecraft in 1980 and by the Cassini space craft in 2003–2009. A package of scientific instruments in the so-called Huygens lander, developed by the European Space Agency, was landed from Cassini onto the surface of Titan in 2005. Some facts about Titan are summarized in Table 8.3. It is notable that, among the moons in the solar system, Titan has the densest atmosphere, with a pressure 50% higher than the pressure of the earth's atmosphere at the planet surface. The atmosphere consists mainly of nitrogen, again somewhat similar to earth (but there is almost no oxygen). However Titan is very cold (mean temperature about $-180°C$), as expected given the distance of Saturn from the sun. There are several lines of evidence that Titan may have a subsurface ocean, somewhat like Europa, and the main interests of astrobiologists in Titan focus on the possibility that there could be some biological activity in the ocean. An image of the surface of Titan, taken by the Huygens lander, is reproduced in Fig. 8.10.

In summary with regard to Titan, there is no immediate indication of biological activity on Titan and the main hope for finding life there arises because the moon is likely to have a subsurface ocean, possibly containing liquid water though the surface liquids observed on Titan are probably methane. If life is found on Titan, it will be of an "extremophile" type because the environment on or beneath its surface cannot be called habitable by terrestrial standards.

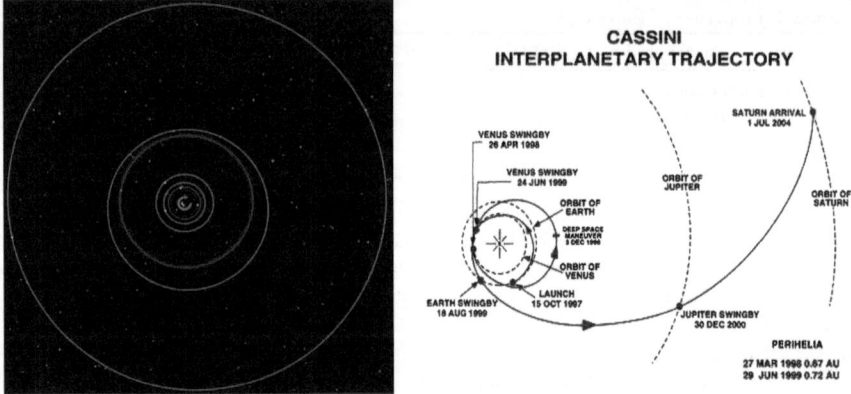

Fig. 8.9 *Left*: Orbits of the moons of Saturn. The orbit of Titan is shown in *red* [6]. The other orbits are those of Saturn moons (from farthest from Saturn to closest to Saturn) Iapetus, Hyperion, Titan, Rhea, Dione, Tethys, Enceladus, Mimas. 53 moons of Saturn have been discovered and named [4]. *Right*: Path of the Cassini space craft [7]

Table 8.3 Properties of Titan [4]

Discovered by Christiaan Huygens on March 25, 1655

Orbital characteristics:
    Semi-major axis: 1,221,870 km
    Eccentricity: 0.0288
    Orbital period: 15.945 days
    Inclination: 0.34854° (to Saturn's equator)
Satellite of Saturn
Physical characteristics:
    Mean radius: 2576 ± 2.00 km (0.404 Earths)
    Surface area: 8.3 × 10^7 km
    Mass: 1.3452 ±0.0002 × 10^{23} kg (0.0225 Earths)
    Mean density: 1.8798 ± 0.0044 g/cm^3
    Equatorial surface gravity: 1.352 m/s^2 (0.14 g)
    Escape velocity: 2.639 km/s
    Rotation period: (synchronous)
    Axial tilt: Zero
    Temperature: 93.7 K
    Surface pressure: 146.7 kPa
    Atmospheric Composition: 98.4% nitrogen, 1.6% methane

8.4 Summary and Outlook for the Search for Life Using Robotic Space Probes

These and other NASA searches have found no signature of life on any planet or planetary moon of our solar system [9]. Though they are not often described this way by NASA and its supporters, these failures have been a great surprise and

Fig. 8.10 Image of the
surface of Titan taken by the
Huygens lander [8]

disappointment to many of the scientists involved. I do not want to imply that the space exploration described here was not scientifically worthwhile. On the contrary it was a scientific and technical triumph which hugely expanded the consciousness of humankind. The fact that the results concerning the presence of life contradicted, to some extent, the prevailing views *increased* the accrued scientific value of the explorations. However, the definition of 'life' implicitly used in these searches has been quite constrained: It assumes a hydrocarbon based biology using water as a solvent and, for some experiments, generating oxygen as a result of metabolic processes. As discussed in Chaps. 4 and 9, we are aware of three broad possibilities concerning the origin of life: First, the basic approach of 'genome first' models may be vindicated by finding a way around the statistical paradox which seems at present to face it. In that case, space searches may find life using their current assumptions, though the non-terrestrial planets and moons do not offer a very fertile ground for life to develop even in that case. Second, the general approach of 'proteins/metabolism first' models may turn out to be correct. In that case, it is likely that the kind of space searches we have described here are not likely to succeed unless the experiments are designed to look more generally for metastable, autocatalytic systems. Thirdly, it may be that terrestrial life is a rare event even if life is defined in a general way. In that case, space searches for life are extremely unlikely to ever succeed.

References

1. //history.nasa.gov/SP-4212/ch3.html
2. //history.nasa.gov/SP-4212/ch3.html
3. http://history.nasa.gov/SP-4212
4. R. Shapiro, *Planetary Dreams* (Wiley, New York, 1999), pp. 207–11
5. NASA/JPL/DLR http://solarsystem.nasa.gov/galileo/
6. From http://en.wikipedia.org/wiki/File:Titanauthors en:User:Rubble pile—Rubble pile]] in Celestia, vectorized byMysid in Inkscape, used under license http://creativecommons.org/licensessa/3.0/deed.en
7. NASA image: http://saturn.jpl.nasa.gov/photos/imagedetails/index.cfm?imageId=776
8. Photo ESA/NASA - University of Arizona by permission
9. For more detail concerning the NASA search for life readable sources are B. Jakosky, *Search for Life on Other Planets*. (Cambridge University Press, Cambridge, 1998) and D. Schulze-Makuch, L.N. Irwin, *Life in the Universe*. (Springer, Berlin, 2004), Chapter 9

Chapter 9
Policy, Ethical and Other Implications

The results of this review of evidence bearing on the probability of extraterrestrial life can be briefly summarized. Our review of the science relevant to estimates of the factors in the Drake equation

$$N_{civ} = N_{gal} f_{star} f_{planet} f_{life}$$

led to values of the first three factors in the range

$$N_{gal} f_{star} f_{planet} = 10^{6 \pm 2}$$

Most of the uncertainty is associated with estimating the likelihood of earthlike planets and estimating the likelihood of hospitable climates on them. Science on both questions is progressing rapidly and the uncertainty can be expected to get smaller in the near future. On the other hand we have discussed estimates of f_{life} in the range

$$10^{-300,000} < f_{life} < 1$$

One may say that such a range means that we know nothing whatever about f_{life}. It arises almost entirely from our ignorance of the prebiotic origin of life and not from uncertainties (which exist but are much smaller) in our understanding of the process of species evolution after life gets started. Our discussion has suggested that estimates giving the high end of the cited range for f_{life} usually use a logically fallacious argument based on the astronomically short time which was required for life to appear on earth. Thus we suggest that the high end of the range for f_{life} is substantially less than 1 but this leaves a very large range of uncertainty. We must also bear in mind that even this very tentative conclusion about f_{life} is very sensitive to the definition which is used in concluding that a given dynamical system is 'life' or a 'biosphere'.

With regard to the 'top down' estimates based on failure of efforts to directly observe extraterrestrial life, we summarize as follows:

J. W. Halley, *How Likely is Extraterrestrial Life?*, SpringerBriefs in Astronomy, DOI: 10.1007/978-3-642-22754-7_9, © The Author(s) 2012

We deduced that the failure to observe colonization of earth by extraterrestrial civilizations (assuming for reasons we discussed that UFO reports do not represent evidence of any such colonization) imply that very long lived civilizations which travel rapidly between stars and survive tens of millions of years, probably do not exist in the galaxy. The failure of SETI electromagnetic all sky searches to observe signals attributed to extraterrestrial intelligence was interpreted to mean that the existence of more than 10 biospheres in the galaxy emitting signals of the sort sought is extremely unlikely. The failure of NASA planetary probes and similar ones launched by other nations to find life is consistent with scenarios in which life is very rare in the galaxy, but it does not put significant new limits on the probability of extraterrestrial life.

In the light of this review, three possible types of answers to the title question have emerged. Each appears to have an appreciable probability of being correct. It is interesting that these three possible answers would have rather different implications for both the meaning and the short term collective agenda of the human enterprise. To review once more, the three open options appear to be

1. Life emerges quite readily on habitable planets leading to biochemically similar life in each instance. This is the dominant, though far from unanimous, view of biologists and astrophysicists. It often takes the form of 'genome first' models for the origin of life. It assumes that problems with those models are not fundamental and will be resolved by technical improvements. This view tends to have problems with various versions of the Fermi paradox, unless difficulties with postbiotic evolution prevent biospheres from evolving to a stage in which the biospheres live a few million years or emit signals of the type sought by SETI. The failure to observe, to date, any evidence of even primitive life forms on the planets of the solar system or on meteorites is sometimes explained within this option by models in which evolution after the prebiotic stage usually leads only to subsurface, primitive life. One could call this the 'life is dull' hypothesis. It is the preferred, but not universal, option chosen by scientists.

2. 'Life' emerges readily on habitable planets but only if one broadens the definition of life to include a wide range of metastable systems which a terrestrial biologist (or astrobiologist) might not recognise as life. This is the view of a minority of 'Proteins first' and 'metabolism first' theorists. Redefining life along these lines makes it easier to account for the early appearance of life on earth. The possible unfamiliarity of the emergent chemistries makes it easier to account for the Fermi paradox than it is in case 1. One could say this is the 'life is weird' hypothesis.

3. Life on earth emerged because an improbable event occurred to initiate it. This view is distasteful to most scientists, but it is not unscientific as long as only one biosphere has been observed. It does not encounter any problem with the Fermi paradox because it predicts that no life systems or SETI signals will be found. It is *not* necessary to suppose that supernatural events of any kind are involved in this option. This could be termed the 'life is rare' hypothesis.

A notable feature of the literature on this subject is that, even confining attention to scientifically competent authors, almost all recent accounts of the subject, including those in textbooks, hypothesize one of these options to be true and often dismiss the other two with brief and questionable arguments. For example the excellent astrobiology text by Ulmschneider [1] states in the preface:

> Although no fossil traces of such beings have ever been found, most of us firmly believe that nonhuman intelligent beings do indeed exist. This conviction is derived from the staggering size of the universe with roughly (10^{22}) stars, which makes it inconceivable that we could be the only intelligent society in the universe.

This is a strong statement postulating option 1. It is a legitimate hypothesis, but the evidence we have cited, while consistent with option 1, is very far from showing that it is preferrable to options 2 and 3. (Aspects of the statement are easily criticized: We have emphasized that the size and age of the universe are not large enough allow an easy demonstration that earthlike biospheres will frequently assemble themselves. And many rational and competent scientists have 'conceived' that 'nonhuman intelligent beings' might not exist in the observable universe, so such a position is not 'inconceivable', though of course it may not be correct.) The rest of Ref. [1] makes essentially no mention of the possibility of options 2 and 3, while describing the information available to the scientific program implied by option 1 with admirable thoroughness.

On the other hand a somewhat more even handed popular [2] treatment by a very competent molecular biologist is entirely devoted to convincing a popular audience that option 2 is to be preferred, at least psychologically. The proponents of option 3 are dismissed as 'Sour Lemons', which is not a term very conducive to rational debate. Some of the psychological issues which may be at play here are briefly discussed below, but Ref. [2] does not make a very sharp distinction between preference as in 'true' and preference as in 'psychologically satisfying'. Option 1 is also dismissed in Ref. [2], more professionally, by detailed and convincing, if selective, criticism of some of the scientific work reported in pursuit of the program involved in option 1. But while Robert Shapiro has contributed tremendously, in my view, to the conception of option 2, this very partisan attempt to tip the balance of public opinion toward option 2 cannot be regarded as scientifically justified. Option 2 is the most interesting option by far, and some individuals find option 3 'dismal', but whether option 2 is the most likely one is simply unknown. Other books on the subject by eminent scientists take an equally unyielding position in favor of option 3. Here the matter is somewhat more complicated because advocates of option 3 sometimes have some kind of quasi or explicitly religious agenda which motivates them to prefer option 3. However the reason that option 3 is listed here is not associated in any way with those motivations. The factual evidence is consistent with option 3, and indeed the evidence fits option 3 rather more easily than it fits the other options, though what we have reviewed here cannot be said, in my view, to establish it as preferred. Upon reading such books, I am reminded of an oft quoted statement by Scott FitzGerald: "the test of a first-rate intelligence is the ability to hold two opposed ideas in the mind at the same time, and still retain the ability to function". Whatever the other failings of that author may be, this seems a very good description of a requirement for good scientific work. It is doubtful that many of the authors on the subject of the likelihood of extraterrestrial life have managed it very well.

If one believes, as I do on the basis of the data and arguments that have been presented, that each of the three possibilities has, at present, a significant chance of being right, then it is appropriate to consider the policy and philosophical implications of each of them.

9.1 Short Term Policy Implications

The most direct policy implications concern possible directions for the searches for extraterrestrial life itself. Because option 1 above is the dominant view, both the SETI, electromagnetic and the NASA space probe efforts are mainly focussed on search strategies that assume a large degree of similarity between the sought biospheres and our own. If the other two options are of comparable likelihood, our review suggests more effort (in light of option 2) to broaden searches so that they seek more general forms of autocatalytic metastability. We have made some explicit suggestions for such revised search directions in Chaps. 4 and 6. Option 3, regarded as a falsifiable hypothesis is continually tested by all kinds of SETI and space probe studies, as is appropriate. The results of such studies, however, are not always analysed in this way, as, for example in the case of the seti@home results discussed in Chap. 7 and such analysis is to be recommended.

9.2 Implications for Environmental Ethics

If the consensus option 1 turned out to be correct, then it would be appropriate to regard the terrestrial biosphere as a kind of frequently repeated experiment. If the terrestrial version of the experiment failed there would be, of course, regrettable implications for the species and communities it housed, but from a more universal perspective one would have no particular cause for concern. Another run at evolution might turn out better and in any case would be interesting. The moral and ethical implications of the existence of failed biospheres would in that case be similar to the moral and ethical implications of the existence of extinct species in our own biosphere: extinction is to be avoided and regretted but it is inevitable and not really tragic because other species arise through evolutionary processes. Perhaps partly because option 1 is a dominant view, a lot of public policy and many private attitudes are consistent with such a view, which is also encouraged by religious traditions which regard earthly existence as transient, morally tainted and somehow not as real as other imagined existences. Though the origins of human attitudes and actions are obviously complex and heavily affected by a very poorly understood genetic heritage, the view of terrestrial life as a transient experiment of no huge worth is consistent with a cavalier attitude toward war, genocide, starvation, disease, species extinction and environmental destruction. Such cavalier attitudes are quite common. Sometimes the connection is quite explicit, as when fundamentalist religionists welcome destruction as punishment for sin or as opportunity to participate in another existence.

If option 2 turned out to be correct, then the logic just described for option 1 would in principle still apply. But if alternative biospheres were on average as different from the terrestrial one as, for example, the Great Red Spot on Jupiter (which I do not mean to seriously insist is alive) then humans might have considerably more difficulty regarding the prospective destruction of their biosphere with as much equanimity as they would in the case of option 1. If alternative biospheres are very alien, then they might not not be regarded as merely interesting extensions or reruns of biological evolution on earth. It seems at least possible that the collective will to protect the biosphere in the broadest sense would be stronger than it is now if people believed that option 2 were correct.

In the case of option 3 there is essentially no possibility of taking comfort in the idea that the extinction of our biosphere would be merely part of a continuous round of natural experiments. If humans value the complexity, consciousness and other experiences associated with our biosphere, then in the case of option 3, the only possibility for assuring the continued existence of such phenomena and experiences anywhere in the observable universe would be to preserve the existing biosphere. Taken seriously, this would imply a radically more serious concern about long term biosphere protection, broadly conceived, than is evident in much of the contemporary human population. For example, acceptance of the rareness of life in the universe could have implications for public policies associated with prevention of devastating wars and environmental destruction if policy makers believed that they were responsible for the maintenance of an extremely rare phenomenon in the cosmos.

9.3 Implications for Views of the Human Condition

If life occurs often (option 1), the failure to observe very long lived (10s of millions of years) rapidly migrating civilizations suggests that option 1 would have the implication that very advanced civilizations, presumably arising frequently because biospheres arise frequently in this option, are always either short lived or do not choose to migrate. This in turn might suggest poor prospects for the long term future of the earth's civilization. (We discussed other arguments in Chap. 6 suggesting that the probability of a long lifetime for the earth's biosphere is low.) This implication of option 1 has been noted by Bostrom [3] and others.

Option 2 is closest to the old pluralist view taken by many scientists in the Eighteenth and Nineteenth centuries. It removes any sense of specialness about our biosphere. This view is the most appealing one to materialistic scientists.

In option 3, the implied uniqueness of earthly life in the observable universe would revive a sense that the biosphere inhabits a unique place in observable universe. Though this a scientific hypothesis, consistent with the known facts and not derived from religious hypotheses, it echoes aspects of the position taken by the churches in the eighteenth and nineteenth centuries and differs significantly from the pluralist position often taken by scientists then.

9.4 Psychological Implications

Option 1 implies biology as usual on other stars and planets. In this option not much psychological adjustment is required. This may account in part for its widespread acceptance among scientists. The second option implies very exciting and disturbing possibilities and responses reminiscent of the emotions evoked by alien horror show movies. It appeals psychologically to more speculatively inclined scientists and other imaginative people with a taste for fantasy. Psychological reactions to option 3 include relief and a sense of affirmation of ancient prejudices but also feelings of isolation and pointlessness. It is unclear that the latter are any more warranted than the former. Psychologically, some individuals are exhilirated by 'wide open spaces' while others are repelled by them, preferring crowded cities and regarding deserts and open landscape as 'dismal'. For example in [4] one finds reference in the context of Option 3 to a 'magnificent destiny' for mankind because it would mean that the universe is open to exploitation and settlement by humans or their successors. Other authors such as Shapiro are psychologically repelled by the isolation implied by Option 3. Psychological resistance to option 3 also arises because it conflicts with the scientific pluralist tradition which has found over several centuries that there is nothing unique about the particular situation of the earth in the universe. However this nonuniqueness is based on evidence and is not a general scientific principle (though it is sometime cited [1] as a reason for rejecting Option 3). Psychologically, Option 3 might actually have positive features from the point of view of the human future. There was never any assurance that if we did encounter extraterrestrial civilizations they would turn out to be benign or beneficial to humans. Indeed if one is to extrapolate from human experience, such encounters might be predicted to have a high probability of being devastating. Among many examples one recalls the well documented fate of civilizations in Mexico and South America after invasion by Europeans. Isolation may be preferable. It is not evident that life becomes less valuable, from whatever point of view, as a consequence of being rare. Usually, in fact, goods which are rare are regarded as more valuable than if they are common.

9.5 Implications for the Long Term Future

We have confined attention to the likelihood that humans will observe other biospheres in the near future. The standard description of the universe outlined in Chap. 1 can be used to produce a prediction for the long term future up to 10^{14} years and beyond [5]. These extrapolations should be treated with caution because they assume that the laws of physics and the model of the universe which fits currently known facts can be used at very long times and large distance scales for which the laws have not been empirically tested. (In fact, Ref. [5] was written before the unpredicted acceleration of the expansion of the universe had been observed.) Nevertheless it is of some interest to note that stars and planets, which are the essential features required for the kind of biospheres we have been postulating, are predicted to survive about 10^{14} years before the expansion causes the stars to extinguish. That is about 10^4 times the current age of the universe. In options 1 and 2, this would certainly lead to a profusion of biospheres. In option 3 the possibility arises that if, as assumed in that option and discussed in Chap. 4, the average time $t_{average}$ for a biosphere to appear on a habitable planet is much longer than 10^{10} years, it might nevertheless be less than 10^{14} years. Thus if it turned out that $10^{10} years < t_{average} < 10^{14} years$ one could anticipate crossing over from option 3 to option 1 or 2 in a time in the far future when stars, planets and biospheres could still exist. Visiting the resulting biospheres would be even more daunting than it seems now because they would be farther apart, but they could still be observed. However another consideration enters: If $t_{average} >> 10^{10}$ years then the required time is longer than the lifetime of stars on the main sequence, like the sun. If biospheres required, on average, such a long time to develop they could only appear in abundance in the future on planets orbiting low mass, long lived stars (of stellar type K5) which live about 7×10^{16} years [6]. If one conjectures (very optimistically) that a human or post-human civilization survived that long, then it could only do so by migrating to another star, since the sun's lifetime is of order 10^{10} years.

9.6 Can a Probability Distribution for a Rare Event be Defined in the Case of Option 3?

The entire exercise of estimating a probabilistic answer to the question of the likelihood of extraterrestrial life may be challenged, particularly in the case of option 3. There is only one universe, therefore there is no ensemble of cases from which to determine a probability distribution and a probability cannot be defined. (It should be emphasized that when I say universe here and elsewhere, I mean the observable universe, containing about 10^{22} stars in 10^{11} galaxies. The size of the observable universe is defined by the distance which light has been able to travel in the age of the universe. Nothing can ever be known experimentally about the nature of any aspects of the universe at greater distances than that, and I have not discussed them. If one speculates that the universe beyond the limits of observation is infinite, then repetition of our biosphere and variants on it could presumably be shown to occur somewhere with unit probability. But this is observationally irrelevant.)

There are several ways to imagine an ensemble in option 3:

In some cosmological models [7], there are many universes. In other cosmological models the universe retracts and expands many times. In these and perhaps other models, there are many instances of an expanding universe and one can speak of the probability that one of them contains one or more 'civilizations' like ours. Unfortunately, universes in which there are no observers cannot be confirmed to exist directly by ordinary scientific methods, so any confirmation of the correctness of such models will have to rely on indirect evidence. Thus the validity of the objection that probabilistic methods cannot be used here is not really settled until we find a way to test these many universe models. (A similar, though somewhat more subtle, problem arises when one attempts to apply the concepts of quantum mechanics to the universe as a whole.)

If we allowed repeated instances of the big bang initiation of the universe to be part of the argument, either because we could compute what would happen if they occurred or because we had a falsifiable model in which such repeated initiations have or are occurring, then a concrete meaning could be associated with the probabilities obtained from the Drake equation, even if those probabilities were very low. For example, if f_{life} were so small that $N_{civ}/N_{gal} < 10^{-13}$ (which seems possible in the random polymer assembly model) then the probability of finding even 1 civilization in the observable universe would be less than 1. Let the number of galaxies in the observable universe be N'_{gal} (N'_{gal} is, presumably coincidentally, of roughly the same order of magnitude the number of stars N_{gal} in a galaxy). The probability of observing a civilization in a given universe is then $N_{civ}N'_{gal}/N_{gal}$ if that number is less than 1. Then one will find such a civilization once in every $N_{gal}/N_{civ}N'_{gal}$ universes, formed by repeating the big bang synthesis under the same macroscopic conditions but with different microscopic conditions, so that the laws of physics in each universe are the same but the microscopic initial conditions imposed by the big bang are different, in each case. Note that these repeated instances of universe formation with the same laws but different initial conditions are different from those envisioned by string theorists who imagine that the laws of physics are different in each universe.

9.7 Anthropic Principles, Religion and Other Nonscientific Aspects

Throughout this discussion, a strong effort has been made to stick to conventional scientific methods, reasoning and established facts. However it is well known that a great deal of speculation on the question of the likelihood of extraterrestrial life has been strongly influenced by a variety of considerations which do not fall within these scientific constraints. Here I briefly review some of these, partly in order to distinguish them as clearly as possible from the arguments being made elsewhere in this volume.

9.8 Biases Held by Scientists

The most weakly unscientific form of thinking which sometimes enters consideration of this subject is a perceived tendency by otherwise very competent scientists to make inappropriately anthropocentric hypotheses in their work on this subject. We have discussed two examples of this at some length.

Many scientists have assumed that $f_{life} = 1$ on the basis of the patently incorrect argument that the fact that the appearance of life on earth was rapid on an astronomical time scale is evidence that the probability of that event was high. It is quite odd that so many otherwise competent professionals have fallen into this elementary error and it may suggest a kind of unconscious bias.

A second example is the choice of a very anthropocentric algorithm for determining whether electromagnetic signals are emanating from complex extraterrestrial biospheres. Basalla [8], among others, has suggested that this also suggests an unscientific bias.

Both examples, and others which could be cited, suggest that many scientists badly want to believe that extraterrestrial life or extraterrestrial civilization much like our own is abundant, and that they have sometimes allowed such desires to unscientifically influence their scientific work.

The origin of these biases may be based in the history of science during the last few centuries: During that time the conflict between religion and science in most cases found the scientific community espousing, and presenting strong evidence for, a universalist, nonanthropocentric view of nature and the universe. For example strong scientific evidence for the Copernican view of the solar system and for the universal application of the same physical laws operating throughout the physical universe has appropriately repudiated previous medieval views closely tied to religion.

But historically, the scientific community also often speculatively extended this very successful idea that there is nothing special about our particular planet to the issue of the likelihood of extraterrestrial life. The reasoning goes: If there is nothing special about our position in space or about our physical laws, then there is not likely to be anything particularly special about our biosphere of living beings either. The successful demonstration that biological systems are entirely subject to physical laws reinforces this view. The attitude has been associated with names like 'pluralist' and 'the principle of mediocrity'.

However the evidence I have reviewed may suggest that, although the physical laws are indeed universal and there is nothing very special about the earth's partic ular location or the nature of our planet in the universe, one thing about our planet, namely its biosphere, may be extremely unusual (Option 3). (Please note that this would not concede anything to the 'intelligent designers' briefly discussed below.) Such a special role for the earth is contrary to the general spirit of several centuries of scientific discovery, though I believe the arguments for it, while not conclusive, are scientifically sound. However Option 3 makes many scientists profoundly uncomfortable, possibly because of this history, and I think this may account for some of the mistakes and anthropomorphic biases which have occurred in the scientific work on the question.

9.9 The Anthropic Principle

I next turn to a line of argument which is close to a scientific one, while straying from the conventional definitions of science as I will discuss. In this approach one asserts a so called 'anthropic principle' which roughly states that predictions concerning what we will observe in the universe should, as a kind of constraint on the theories used to make the predictions, take account of the fact that we humans exist in the universe and are making the observations [9].

In fact most of the theories of physics do not take explicit account of the existence of observers (though it can be argued that some formulations of quantum mechanics do). The original suggestions of an anthropic principle arose because physicists [10] noticed that slight changes in some of the fundamental parameters of established physical theories would yield alternative universes in which galaxies, stars and chemistry as we know it would not occur. Whether universes without galaxies, stars and chemistry as we know it would be capable of evolving some form of metastable autocatalytic activity is usually not discussed.

More recently, some very speculative, but popular, theories concerning the fundamental nature of matter on extremely small scales suggested that a huge number of alternative universes might be possible (or might exist) and some physicists suggested using the anthropic principle to explain why we are observing a universe in which life (and observers) is possible [11]. This reasoning is widely but not universally regarded as unscientific in the physics community, but it has been seriously put forward by some very competent physicists.

In the present volume I have consistently taken the physical laws to be those which have been confirmed by experiment and I have not worried about what would happen if those laws were different (as the theorists cited above do). What we found by proceeding in that way is that, while life and observers are (obviously because we exist) possible, they may not be at all likely. Thus our exploration suggests that the existing physical laws may *just barely* allow the existence of life and observers.

This illustrates that the 'anthropic principle' as stated above would need to be stated more precisely if it were to be useful as a predictive principle. For example, one might postulate that the laws of physics must be such that one set of observers occurs in the universe with probability one. But we have seen that with our physical laws it seems quite possible that, even though we undoubtedly exist, the probability of the occurence of a biosphere like ours in our universe may be less than one and our existence may be a rare event (Option 3). Thus an 'anthropic principle' postulate that the laws of physics must be such that one set observers occurs in the universe with probability one is not self evident and may be inconsistent with the physical laws which we experience.

9.10 Intelligent Design

Proceeding from accepted science to progressively less scientific approaches to the issues discussed here, we move next from discussion of the anthropic principle (accepted by a few respectable scientists but not by most) to arguments from 'intelligent design' (also espoused by a few scientists but generally rejected by the consensus of the scientific community).

Individuals and groups who espouse 'intelligent design' arguments range from quite well trained scientists to people with a transparently antiscientific religious agenda and little knowledge of science. The entire subject is embroiled in sociological and political controversy, much of which is irrelevant to the discussion here. I will try to reproduce and comment on the better informed and less obviously biased versions of the position of people espousing 'intelligent design' as a relevant idea in this context.

Essentially, the best informed of them usually focus on the difficulties we have discussed in understanding how models such as the random polymer assembly model for prebiotic evolution could account for a value of f_{prebio} which resulted in the existence of a biosphere such as ours with a probability near 1 (either for the galaxy or the universe). The general line of argument is that, if the probability derived using physical laws is low, then, because we and other entities in the biosphere exist, some agent (an 'intelligent designer') must have intervened to cause the physical laws to be violated.

This argument contains several fallacies. The most elementary one, shared by many scientists who have discussed this problem, is to regard our existence as evidence that the probability of the occurence of a biosphere in the universe is near one. But as I have repeatedly stressed, one data point cannot be used to estimate the probability of an event which is dependent on stochastic processes governed by a probability distribution. Saying that our existence establishes that the probability of a biosphere in the universe is equal to or near one is logically similar to tossing a coin once, getting heads, and estimating the probability of getting heads on subsequent tosses of the same coin as $1/1 = 1$. The probability of the existence of a biosphere such as ours can be less than one (the 'rare event' option 3 possiblity discussed above and in Chap. 4) even though we undoubtedly exist. No contradiction arises and there is no need to invoke any kind of superscientific designer or design principle to explain our existence.

But, there is also a possiblity, also discussed above, that our understanding of prebiotic chemistry is insufficiently refined and that a more sophisticated model than the random polymer assembly model or a 'metabolism (or proteins) first' model as in options 1 or 2 will be established which will show that f_{prebio} is large enough to allow our existence to be an expected event with probability close to one in the observable universe. In that case, SETI searches will have a finite probability of success. Again, no contradiction with existing physical laws arises as a result of our existence. The main difference in the second case is that other civilizations would also be likely to be observed. No need for an agent violating physical law occurs in this case either.

Finally one should mention that I have selected the most sensible version of the intelligent design argument for discussion. Unfortunately, several 'intelligent designers' also attempt to apply similar arguments to the evolutionary factor f_{evol} in f_{life}. As discussed above, there is abundant experimental evidence that once unicellular organisms form, natural selection results in very rapid evolution of species. This has occurred repeatedly in the paleontological record after mass extinctions and, on a much more rapid time scale, among microorganisms in our laboratories and hospitals. The attempts to apply intelligent design arguments to species evolution after the prebiotic stage are totally without scientific merit and seem to be mainly motivated by sociological and political considerations.

9.11 Religious Attitudes

Finally, I will very briefly discuss the attitudes of religious thinkers toward these issues, though I have explicitly excluded such considerations from the arguments presented so far. For a variety of reasons, the subject attracts the interest of people interested in religion. For one thing, most religions include ancient creation myths in their doctrines, and reconciling those myths with modern scientific knowledge has often proved troubling. Indeed literal interpretation of most such myths, if treated as scientific hypotheses in the sense of Popper, have been as thoroughly falsified by scientific facts discovered in the last two centuries as any theory has ever been falsfied. Nevertheless, these ideas continue to survive in the population. In the seventeenth and eighteenth centuries when the untenable nature of these myths was less evident, the discussion of the existence of extraterrestrial life was often based on religious precepts as reviewed in Ref. [12].

As data concerning the number of stars in the sky and the similarity of the stars to our sun became known, scientists began speculating on the presence of life around other stars. One argument which often occurred was that, if, as was usually supposed, a supernatural creator had produced life on earth, then it would be a waste of the other stars if that same creator had not produced life on planets near them as well. Others, however, argued that multiple life systems, particularly involving intelligent beings would contradict religious doctrines implying the uniqueness of humans in the universe.

These discussions have a kind of echo in the arguments presented here, though, in the more informed versions of modern discussion, the supernatural creator has usually disappeared. Physicists and others who have supposed in the recent past that $f_{life} \approx 1$ are reaching conclusions somewhat like those of their eighteenth century predecessors who argued that God would not waste the stars by not producing life on them. (Though I have argued that the arguments which those contemporary physicists use are based on a statistical fallacy, they do not, on the other hand, invoke a Creator in the argument.)

Those who argued in past centuries that God only created one intelligent species were reaching conclusions somewhat like the ones which I (in Option 3) and others such as Hart have been suggesting, though on the basis of entirely different arguments. In present day religious communities the most vocal participants in the discussion seem to think that the discovery of extraterrestrial intelligence and even of extraterrestrial unicellular life would be theologically disturbing.

My primary objective in mentioning these matters is to disassociate myself as thoroughly and clearly as possible from the suggestion that Option 3 has any theological implications. The arguments I have presented here have been as scientific and objective as I could make them and I hold no explicit or implicit assumptions based on extrascientific hypotheses concerning the likelihood of extraterrestrial life.

Another religious overtone in the search for extraterrestrial life arises because many ancient myths suggested that superhuman beings exist in the 'heavens', defined in various ways as scientific knowledge grew. Remanents of this belief may unconsciously affect some contemporary attitudes. Some people have on this basis metaphorically equated a search for extraterrestrials with a 'search for God'. Using our findings particularly from Chap. 6, one can argue that the absence of colonization makes the existence of some kinds of god-like civilizations millions of years old extremely unlikely on the basis of current evidence. From that point of view, the scientific information assembled here does appear to falsify the hypothetical existence of certain kinds of superhuman gods. It is quite possible that this is a good thing from the human point view, because there has never been the slightest indication that such a superhuman civilization, if it existed, would be benevolent toward the earth and its inhabitants.

References

1. P. Ulmschneider, *Intelligent Life in the Universe: Principles and Requirements Behind Its Emergence* (Springer, Berlin, 2006)
2. R. Shapiro, *Planetary Dreams: The Quest to Discover Life Beyond Earth* (Wiley, New York, 1999)
3. N. Bostrom, Technol. Rev. May/June 72 (2008)
4. R.N. Bracewell, in *Extraterrestrials, Where Are They?* (Cambridge, 1995) p. 39
5. F. Adams, G. Laughlin, Rev. Mod. Phys. **69**, 337 (1997), A different view of the long term future appears in Freeman Dyson. Rev. Mod. Phys. **51**, 447–460 (1979)
6. I.S. Shklovskii, C. Sagan, *Intelligent Life in the Universe* (Table II Holden-Day, San Francisco, 1968)
7. A. Linde, J. Cosmol. Astropart. Phys. 2007, 1475 (2007)
8. G. Basalla, *Civilized Life in the Universe: Scientists on Intelligent Extraterrestrials* (Oxford University Press, Oxford, 2006)
9. J.D. Barrow, F.J. Tipler, *The Anthropic Cosmological Principle*, (Oxford University Press, Oxford, 1986), Many scientists regard Tipler's later work on this topic as unscientifically theological
10. For a relatively recent review: C.J. Hogan, Rev. Mod. Phys. **72**, 1149–1161 (2000)
11. L. Susskind, The anthropic landscape of string theory, preprint hep-th/0302219 (2003)
12. M. Crowe, *The Extraterrestrial Life Debate* (Cambridge Press, Cambridge, 1998)

Appendix 1.1
Forms of the Drake Equation

We will use the Drake equation in the form (Eq. 1 in Chap. 1):

$$N_{civ} = N_{gal} f_{star} f_{planet} f_{life}.$$

In most astronomy texts one finds the form as originally written down by Drake (Drake wrote N for N_{civ}):

$$N_{civ} = R^* f_p n_e f_l f_i f_c L$$

in which R^* is the rate at which stars with the right chemistry are born, f_p is the fraction of such stars with planetary systems, and n_e is the number of earth-like (interpreted as habitable) planets per planetary system. L is the average lifetime of civilizations. In this book I will usually use the notation $\tau_d \equiv L$. The product $f_l f_i f_c$ is the fraction of stars with the right chemistry and a habitable planet on which a civilization is found *at some time during the lifetime of the star*. This is a little different from the definition of f_{life} in the Hart form (Eq. 1) which is the fraction of habitable planets on which life is present *when they are observed*. To understand the relation between the Drake form and Eq. 1, multiply and divide the Drake form by the average lifetime τ_{star} of a star (roughly 10^{10} years) with the right chemistry and rearrange slightly:

$$N_{civ} = [R^* \tau_{star}][f_p n_e][f_l f_i f_c (L/\tau_{star})]$$

The square brackets are inserted to highlight the factors corresponding to the factors in the Hart form (Eq. 1 in Chap. 1). In a steady state the birth rate of stars of the right chemistry must equal its death rate so that

$$R^* = N_{gal} f_{star} / \tau_{star}$$

and

$$R^* \tau_{star} = N_{gal} f_{star}$$

J. W. Halley, *How Likely is Extraterrestrial Life?*, SpringerBriefs in Astronomy,
DOI: 10.1007/978-3-642-22754-7, © The Author(s) 2012

The second factor in square brackets corresponds to f_{planet} because the latter takes implicit account of the possibility of more than one habitable planet per planetary system:

$$f_p n_e = f_{planet}$$

The last factor in square brackets is f_{life}

$$f_l f_i f_c (L/\tau_{star}) = f_{life}$$

where the factor L/τ_{star} takes account of the difference between the definition of $f_l f_i f_c$ which gives the probability that a civilization exists at some time during the lifetime of the star and f_{life} which is the probability of a planet harboring a civilization when it is observed. (I avoid saying 'at the present' because for distant objects observations are of events occuring in the distant past.) If one assumes that there is nothing special about the moment of observation, then if a civilization lives at some time period of length L during the star lifetime of length τ_{star} then the probability of observing it at a randomly selected observation time is L/τ_{star}. (I assume here that $L < \tau_{star}$, omitting the possibility, sometimes discussed, that civilizations might leave their stars to avoid dying when the star ends its life as a supenovae, red giant or red dwarf.)

In the book, one finds one other form of the Drake equation:

$$N_{civ} = (\tau_d/\tau_b) N_{gal}$$

in which $1/\tau_b$ is the rate per star of appearance of civilizations and $\tau_d \equiv L$ is the average civilization lifetime. This is derived from the assumption of a steady state in which the birth rate of civilizations equals their death rate, leaving a steady state population:

$$N_{civ}/\tau_d = N_{gal}/\tau_b$$

From the Hart form (Eq. 1 in Chap. 1):

$$f_{star} f_{planet} f_{life} = \tau_d/\tau_b$$

which can be interpreted as equating death rates and birth rates per star. From the Drake equation

$$[f_p n_e][f_l f_i f_c (L/\tau_{star}] = \tau_d/\tau_b$$

or because $\tau_d \equiv L$

$$1/\tau_b = [f_p n_e][f_l f_i f_c/\tau_{star}].$$

Appendix 2.1
The Doppler Shift

We may think of a star as emitting light from atoms in its atmosphere at fixed frequency, determined by the type of atom in the atmosphere. Let the frequency be $1/\tau$ where τ is called the period (a time. For atoms emitting in the visible part of the spectrum it is very short, of order 10^{-16} seconds.). If the star were not moving with respect to us, as observers, then the atoms would emit a peak in the light wave every time τ and, because the light is moving at velocity c, the peaks would be a distance $c\tau$ apart, which would be the observed wavelength of the detected light (Fig. 1a). Now suppose instead that the star is moving away from us with velocity $v < c$. In that case, the atoms in the star are still emitting a peak in the wave every time τ, but now in this one period the source has moved away from us a distance $v\tau$ before the next peak is emitted, whereas the previously emitted peak has moved toward us a distance $c\tau$ so the total distance between the peaks which we observe (which is also the wavelength which we observe) is $c\tau + v\tau$. The light's wavelength which we observe has been 'stretched' by an amount $v\tau$ (Fig. 1b). This is the Doppler shift.

If we call the wavelength shift $\Delta\lambda$ and the unstretched wavelength λ then we have

$$\Delta\lambda = v\tau$$

$$\lambda = c\tau$$

and dividing the first of these equations by the second

$$\Delta\lambda/\lambda = v/c$$

so the velocity of the receding object can be determined from the wavelength.

J. W. Halley, *How Likely is Extraterrestrial Life?*, SpringerBriefs in Astronomy, DOI: 10.1007/978-3-642-22754-7, © The Author(s) 2012

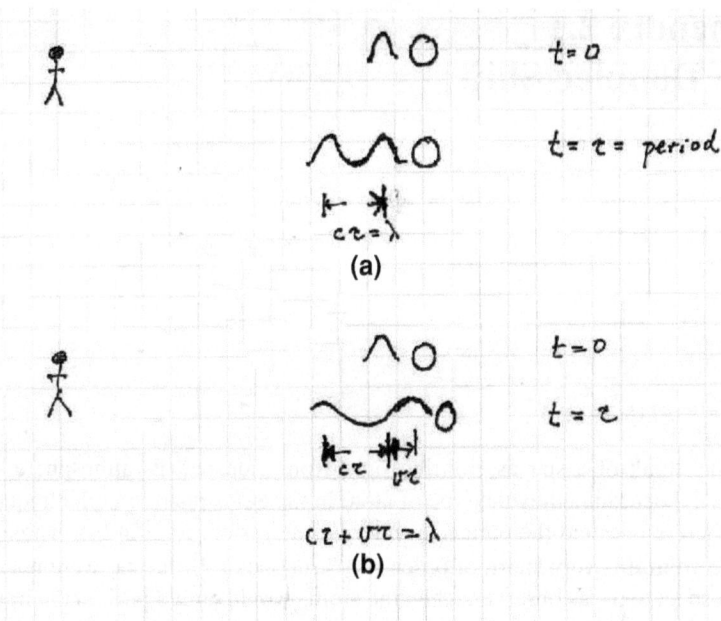

Fig. 1 Sketch of wave forms associated with the Doppler effect

The details of this account of the nature of the Doppler effect only apply to redshifts associated with nearby galaxies and (without restriction) to planets orbiting stars in our galaxy. For more distant, rapidly receding galaxies for which the recession velocity is comparable to the velocity of light, a full treatment requires the use of the general theory of relativity and is not described here.

Appendix 3.1
Doppler Shifts for Circular Planet Orbits

For a single planet of mass M_p orbiting a star of mass M_s with speed v_p at a distance R from the star and with $M_p \ll M_p$ the equation of motion in a circular orbit reduces to

$$M_p v_p^2 / R = GM_p M_s / R^2.$$

Solving for the planet speed gives $v_p = \sqrt{GM_s/R}$. The planet and the star are orbiting the center of mass so, if v_s is the speed of the star,

$$M_p v_p = M_s v_s$$

and

$$v_s = (M_p/M_s)\sqrt{GM_s/R} = M_p\sqrt{G/(RM_s)}$$

On the other hand, $v_p = 2\pi R/T$ where T is the period so

$$(2\pi/T)^2 = GM_s/R^3$$

or

$$T = \sqrt{2\pi R^3/GM_s}$$

which is Kepler's law. In the Doppler shift measurement one measures $v_s \sin i$ where i is the angle of inclination of the orbit to the line of sight:

$$v_s \sin i = (M_p \sin i)\sqrt{G/(RM_s)}$$

One easily solves the last two equations, which give the two measured variables of planetary interest $M_p \sin i$ and R in terms of the measured variables T and $M_p \sin i$:

J. W. Halley, *How Likely is Extraterrestrial Life?*, SpringerBriefs in Astronomy,
DOI: 10.1007/978-3-642-22754-7, © The Author(s) 2012

$$R = T^{2/3}(GM_s/2\pi)^{1/3}$$

$$M_p \sin i = v_s \sin i (RM_s^2/\sqrt{2\pi G})^{1/3}$$

The mass of the star can usually be inferred from its luminosity and color, together with existing models of stellar evolution, so R and $M_p \sin i$ could be calculated if the orbit were circular. In the case of circular orbits, the time dependence of the observed Doppler shift as shown, for example in the figure on p. 19, will be sinusoidal with period T and amplitude $v_s \sin i$. In the more general case of elliptical orbits, one finds an equation for the time dependence of the $v_s \sin i(t)$ in terms of the orbital parameters and fits the orbital parameters to match the observed time dependence. Systems showing more than one period in the time dependence are usually multiplanetary systems and a similar procedure can be used to extract characteristics of the orbiting planets in those cases as well.

Appendix 4.1
Catalysis at Surfaces

One can get a rough understanding of how surfaces enhance reaction rates by thinking about trying to play pool, which is played with balls rolling on a two dimensional table, in a three dimensional environment, for example by trying to trying to hit one tennis ball in the air with another. It is easier to hit one ball with another on a table than it is to hit one ball with another in the air because there are more ways to miss in three dimensions than in three (though the added complication in our sports comparison is that gravity makes it impossible to make the balls stand still in three dimensions.). One can roughly quantify this notion with a simple mathematical model.

Consider the case of a reaction in which a molecule A of radius a interacts with another molecule A to form B. Suppose first that the molecules A are in a solution with concentration D_3 molecules per cm^3 and that they are moving around randomly with average speed v. Assuming that the molecules move in straight lines between collisions. Treat the molecules as spheres of radius a. We will show that the rate at which they run into each other is $4\pi a^2 D_3 v$: as a molecule moves through space, it will hit any other molecule A whose center passes within 2a of it. Thus in time t it sweeps out a cylinder of radius 2a and length vt (Fig. 2). Any molecules inside the cylinder gets hit. The number of molecules in the cylinder is $D_3 \times$ the volume of the cylinder. When the time t is the average collision time, which you want, the average number of molecules in the cylinder should be one. The rate is one over the average collision time: $rate = 1/t = 4\pi a^2 D_3 v$.

Next, suppose that the molecules are attached to a surface with surface density D_2 molecules per cm^2 but that they move freely around on the surface with the same average speed v. A very similar consideration gives the rate as $4avD_2$. This time you should think of the molecules as sweeping out a rectangle of width 4a and length vt.

J. W. Halley, *How Likely is Extraterrestrial Life?*, SpringerBriefs in Astronomy, DOI: 10.1007/978-3-642-22754-7, © The Author(s) 2012

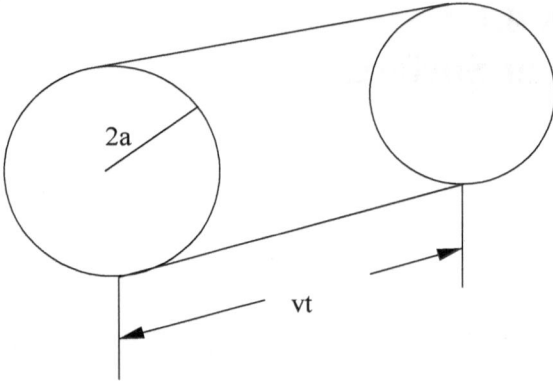

Fig. 2 Illustrating the calculation of the collision rate

The ratio of the two rates is D_2/aD_3 If the distance between molecules is the same in the three dimensional system as it is in the two dimensional system then this is $\gg 1$ as long as the mean distance between molecules is much larger than their size. In fact this is almost always the case. However another factor often further increases the rate at surfaces, namely that the mean distance is smaller on the surfaces because the molecules are attracted to the surface.

Appendix 4.2
A Kauffman-Like Model

To illustrate some issues discussed in the text, I give here some computational results on a Kauffman-like model for the origin of life. Details will be published elsewhere (I. Fedorov and J. W. Halley, unpublished). The model consists of polymers, which one can regard as primitive nucleic acids or proteins. The number of possible monomers is only two, unlike nucleic acids (with 4) or proteins (with 20). The polymers are then represented simply by strings of 0's and 1's: 001100 for example. The reactions in the model are splitting, $001100 \rightarrow 001 + 100$ for example and condensation $001 + 100 \rightarrow 001100$. Each reaction is catalysed by another polymer and cannot take place without that other polymer. The allowed reactions are chosen at random from all the possible ones, starting with a small set of short, starting polymers and building out to reactions involving larger polymers such that each reaction has a probability p of being included. The resulting reaction network is 'grown' for a certain number of steps N_t which is another parameter of the model. Once the network is established in this way, actual reaction rates for each included reaction are also assigned at random, by assigning a random number to each one. Each reaction is assigned a 'forward' rate and a 'reverse' rate with a fixed ratio of the forward to the reverse rate.

The resulting model can be 'run' on a computer to simulate a form of primitive biochemistry, starting with a set of small polymers, termed the 'food set'. The number of polymers of type n_α then changes as the reactions take place. The food set is replenished and the polymers are removed at random to keep the total number of molecules less than a fixed maximum in the steady state. The results are different for each computer 'run' because only the probabiity of each reaction is fixed by the model. Whether, in a given run, a particular reaction takes placed is determined at random, consistent with the assigned probability for that reaction. This means that, in the model, we can reproduce something like what would happen on different planets during prebiotic evolution by running the model with different sets of random numbers.

The distribution of polymers during each run always reaches a 'steady state'. To interpret the results we have to decide which final steady states to regard as lifelike. In the model described here we apply the criterion described qualitatively

J. W. Halley, *How Likely is Extraterrestrial Life?*, SpringerBriefs in Astronomy, DOI: 10.1007/978-3-642-22754-7, © The Author(s) 2012

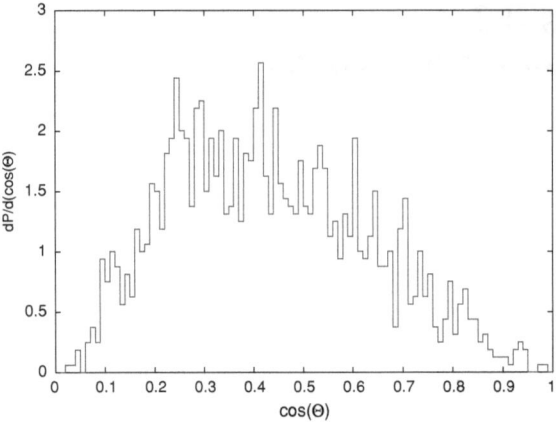

Fig. 3 Distribution of angles between life like final states in a model of prebiotic evolution (*blue*) and for randomly selected states (*red*)

in the text by requiring that the information in the steady state (described in the model by the negative of a 'configurational entropy' of the collection of polymers) be larger than the (low) equilibrium value. (The life-like entropy is required to be low. The lifelike information is required to be high.) We find that such states occur only at small values of p, but that if the p is too small the number of life-like states by this criterion again is small so there is an optimum value of p for creating life-like states. We find that many of the 'life-like' states created according to this criterion are 'fixed points' with no dynamics. Static states are not usually regarded as lifelike, so we add the requirement that the states of interest are 'limit cycles' with polymer populations changing in time.

   In this way we obtain a set of 'life-like' states from the model by running the dynamics of the chemical network over and over again. The question we ask next is, are such life-like states similar to one another? In other words, is this primitive prebiotic evolution model 'convergent' so that the life-like states it produces are all very similar? To address this we note that the states of the model are described by sets of numbers $n_1(t), n_2(t), \ldots$ which describe how many of each type of polymer is present at each time t. In practice, to make the calculations feasible, we have chosen networks so that the number of polymer types is in the range 60–90. In the results shown in Fig. 3 the states were described by collections of 82 numbers. We think of these as vectors in a space of 82 dimensions and characterize the difference between two final states as the angle between these two vectors. If the final states are all very similar, the angles will all be near zero. In fact, if the states were chosen completely randomly (without any dynamics or criteria of low entropy and dynamics) then the angles between state vectors would also be small because equal occupancy of all the polymers would be likely. We show some results (plotted as a function of the cosine of the angle between the state vectors, so that if the cosine is 1 the states are similar), in Fig. 3. One sees that, in the model, the prebiotic evolution is not at all convergent. The states are spread over a very wide range.

Appendix 5.1
Evaluating How an Experiment or Observation Affects the Relative Likelihood of Two Competing Theories

Consider two theories A and B which are consistent with previous data and which are constructed so that they can each produce a number for the probability of some event E (an experimental reading or an astronomical observation for example) which will occur in the future. Suppose that those probabilities, which are calculated before the experiment or observation using the theories are denoted $P(E|A)$ and $P(E|B)$. Now suppose that from previous experiments and calculations, the scientists involved have determined that the probability of theory A being correct is $P_{before}(A)$ and the probability of theory B being correct is $P_{before}(B)$ where the subscript is to remind us that these are the probabilities known before the experiment or observation which could produce the result E is done. Now we imagine that the experiment or observation is done, we get the result E and we want to know: How did the result E change our estimate of the likelihood that the two theories are right? In other words what is the probability $P_{after}(A) = P_{before}(A|E)$ where the right hand side means 'the probability that A is right if the result of the experiment or observation is E' and similarly for B. (Philosophers may want a more precise idea of what is meant by 'right' or 'correct' here but I will assume that the intuitive meaning is clear: we mean the probability that the theories will continue to correctly predict results of the class of experiments or observations for which they were designed.) Obviously if A predicted that E would certainly occur ($P(E|A) = 1$) while B predicted that it certainly would not ($P(E|B) = 0$) then the experiment will have established that $P(B) = 0$ (though not that $P(A) = 1$) and B has been falsified. However things are often not so simple because the theories may, and often do, predict probabilities which are between 0 and 1. In that case we use the identities:

$$P_{before}(A, E) = P_{before}(E|A)P_{before}(A) = P_{before}(A|E)P_{before}(E)$$
$$P_{before}(B, E) = P_{before}(E|B)P_{before}(B) = P_{before}(B|E)P_{before}(E)$$

All the probabilities refer to the situation before the experiment. The left most probabilities are the probabilities that the experimental result E obtained and that the theory is true. Now we divide the first equation by the second, cancel out the factor $P_{before}(E)$ and use the definitions of $P_{after}(A), P_{after}(B)$ to give

$$\frac{P_{after}(A)}{P_{after}(B)} = \left(\frac{P_{before}(E|A)}{P_{before}(E|B)}\right)\left(\frac{P_{before}(A)}{P_{before}(B)}\right)$$

The first factor on the right is given by the theories and the second was assumed known before the experiment. Thus if $P_{before}(E|A) > P_{before}(E|B)$ according to the calculations of the theories, then the experiment has established that the probability of A is higher relative to the probability of B than it was before the experiment (and by an amount which can be calculated.) Carter [1], discusses the question (discussed in Chap. 4) of determining whether, on the average, life occurs on habitable planets in times of order t_{earth} using this relation. Let A be the theory that $t_{earth} < \bar{t}$ and B be the theory that $\bar{t} \approx t_{earth}$. Let E be the event, which we have observed, that life appeared on earth in time t_{earth}. Clearly $P_{before}(E|A) \approx 1$. But some observer has to be present to make an observation and only one life system has been observed (by itself) so in theory B it must also be the case that $P(E|B) \approx 1$ when E is the event that the first and so far only life system has been observed by itself. Therefore the observation by us that life appeared on earth in time t_{earth} does not change our estimate of the relative probabilities of theories A and B at all. (This is a rather abstruse way to put the fact that you cannot make any statistical conclusions about an ensemble from one observation, as described in Chap. 4.)

Appendix 6.1
Diffusion and Random Walks

The diffusion model is the mathematical description of the average behavior of a collection of entities performing random walks. A random walk of an object moving on a simple lattice in space is described as follows. At each time τ_{hop} the walker chooses one of six directions (up, down, right, left, forward, back) at random with equal weighting for all six choices and takes a step of length l in that direction. After N steps, corresponding to time $N\tau_{hop}$, the position $\vec{R}$ of the walker can be described by

$$\vec{R}(t) = \vec{\Delta r_1} + \vec{\Delta r_2} + \cdots + \vec{\Delta r_N}$$

where $\vec{\Delta r_i}$ is a vector describing the change of position at the step i. If one imagines taking many, say M such walks for N steps and averages square of the resulting final positions $\vec{R}_1(t), \ldots, \vec{R}_M(t)$ then the result is

$$\langle R^2 \rangle = (1/M)(R_1^2(t) + \cdots + R_M^2(t))$$

which can be written as

$$\langle R^2 \rangle = \langle \Delta r_1^2 \rangle + \cdots + \langle \Delta r_N^2 \rangle + \langle \vec{\Delta r_1} \cdot \vec{\Delta r_2} \rangle + (N(N-1) - 1) \text{ more cross terms}$$

The cross terms are zero on average and the diagonal terms are all the same giving

$$\langle R^2 \rangle = Nl^2 = (N\tau_{hop}/\tau_{hop})l^2 = 6Dt$$

where $D = l^2/(6\tau_{hop})$.

J. W. Halley, *How Likely is Extraterrestrial Life?*, SpringerBriefs in Astronomy, DOI: 10.1007/978-3-642-22754-7, © The Author(s) 2012

Appendix 6.2
Modeling Diffusion, Birth and Death

A more complete description of the local density $n(r)$ in a model of a spherical galaxy in which civilizations diffuse with diffusion constant D, are born at rate $n_s(r)/\tau_b$ and die at rate $n(r)/\tau_d$ is given (in steady state) by the equation

$$D(1/r^2)d/dr(r^2 dn/dr) - n/\tau_d + n_s(r)/\tau_b = 0$$

where $n_s(r)$ is the density of stars. r is the distance from the center of the galaxy. (Generalization to models which are not spherically symmetric is not very difficult.) As in the text, $D = l^2/6\tau_{hop}$ and l is the mean distance between stars. If we suppose that the star density is a constant n_s inside a sphere of radius R and that the density of civilizations goes to zero at $r = R$ and that the density is not infinite at $r = 0$ the equation is straightforward to solve (change the variable to $u = rn$) with result (for $r \leq R$).

$$n(r) = n_s(\tau_d/\tau_b)[1 - (R/r)(\sinh(kr)/\sinh(kR))]$$

where $k = 1/\sqrt{D\tau_d}$. The quantity $\sqrt{D\tau_d}$ is the distance which a civilization diffuses during its lifetime. If $\sqrt{D\tau_d} \ll R$, corresponding to short lifetimes and/or slow diffusion then $n(r)$ reduces to

$$n(r) = n_s(\tau_d/\tau_b), \quad r < R$$

$$n(r) = 0, \quad r = R$$

which is exactly what the Drake equation would give. On the other hand if $\sqrt{D\tau_d} \gg R$ corresponding to long lifetimes and/or fast diffusion, then the result becomes

$$n(r) = \frac{n_s}{6D\tau_b}(R^2 - r^2)$$

in which the death rate $1/\tau_d$ is irrelevant and the local density is entirely determined by the diffusion rate and the birth rate. Here the density at the origin is $n_s(R^2/D\tau_b) = n_s\tau_{fill}/\tau_b$ where $\tau_{fill} = R^2/6D$ as it was defined in the text. So these

J. W. Halley, *How Likely is Extraterrestrial Life?*, SpringerBriefs in Astronomy, DOI: 10.1007/978-3-642-22754-7, © The Author(s) 2012

two limits give approximately the same answers for the density as the two limits in the more qualitative discussion in the text. A graph of some intermediate cases is shown in the figure.

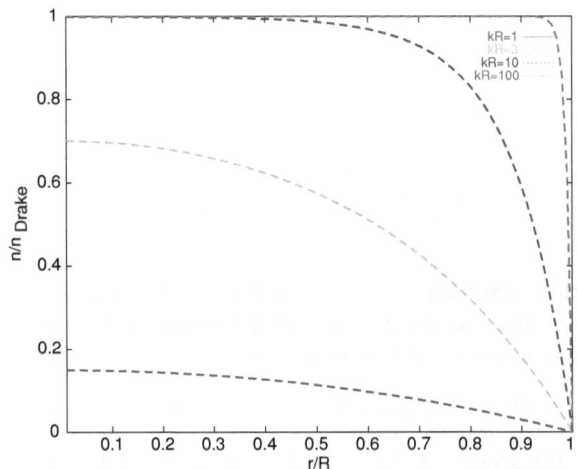

One sees that there is a crossover from one behavior to the other when $kR \approx 10$. In terms of the number N_{gal} of stars in the galaxy we have in this spherical model that

$$kR = \sqrt{6\tau_{hop}/\tau_d} N_{gal}^{1/3}(3/4\pi)^{1/3}$$

and with $N_{gal} = 10^{11}$, diffusion will be relevant when $kR < 10$ giving

$$\tau_d > (1/100)(3/4\pi)^{2/3}(10^{11})^{2/3}\tau_{hop} \approx 2 \times 10^6 \tau_{hop}$$

The hop times are restricted at the lower limit of about 10 years by fundamental physics as discussed in the text. At that limit the crossover from a local density determined by the Drake equation to a local density determined by diffusion will require

$$\tau_d > 2 \times 10^7 \, \text{year}$$

or about 10 million years. This is close to the rough estimate in the text. If hop times as long as 10^4 years are all that can be achieved then diffusion will be irrelevant even if the civilization lifetimes are comparable to the age of the universe. The quantitative estimates will be sensitive to the details of the transport model.

Appendix 6.3
Units of Radiation Dosage

For radiation biology, one starts with the amount of energy deposited in the organism by the radiation in joules per kilogram of organism mass. The unit 1 joule/kilogram is called a Gray, but the more common unit is 10^{-2} Gray = 1 rad. Biological harm is proportional to total energy deposited (not to the rate at which it is deposited). Further, biological harm depends not only on the amount of energy deposited but also on the type of radiation involved. This dependence is taken into account empirically by a 'quality factor' QF which, for relevant radiation sources, is > 1 and is such that the ratios of the harm done by depositions of the same amount of energy from different sources are given by the ratios of their quality factors. Then the 'dose' is expressed in terms of the unit rem defined by 1 rem = QF $\times$ 1 rad. In the US, the legal limit for exposure of the citizenry to radiation above the natural background of about 0.4 rem/year is 0.1 rem/year whereas in the International Space Station the total radiation dose is around 2.8 rem/year. From the US standard one infers that (with a 60 year lifetime) significant harm occurs at more than about 30 rem total lifetime dose, so that astronauts need to limit their total time in space to less than 10 years per lifetime. This is a serious constraint on human interplanetary travel. Very large amounts of radiation shielding would be required in order to allow humans to avoid harm over lifetimes in space. The extremophiles Deinococcus [2], (several reported species) are reported to survive after dosage of 30 KiloGray = 3×10^6 rad or more than 3 million rem.

J. W. Halley, *How Likely is Extraterrestrial Life?*, SpringerBriefs in Astronomy, DOI: 10.1007/978-3-642-22754-7, © The Author(s) 2012

Appendix 7.1
Origin of the 21 cm Line

The 21 cm wavelength radiation arises from a magnetic interaction between the positively charged proton nucleus of the hydrogen atom and the electron orbiting around the nucleus. Each component (electron and proton) acts like a little magnet whose strength is characterised by its so-called magnetic moment. If you think of the magnetism as arising from circulating charges inside the electron and proton and simplify the magnetic model to an intrinsic current loop for each particle, then the magnetic moment would be the current going around the loop times its area. The magnetic moments of the electron and proton have been measured to great accuracy, but here we will be content with a rough estimate. The magnitude of the moments is constrained by the fact, known from study of atomic spectra and formalised in the quantum mechanical description of those spectra, that the smallest possible angular momentum associated with the circulating currents is (of order) Planck's constant h divided by 2π. Now imagine that the model current loops have radii a_e and a_p for the proton and electron, that the charges are $-e$ and $+e$ and that the masses of the particle are m_e and m_p. If the entire mass is circulating (as well as the charge) then the magnetic moments, denoted μ_e and μ_p would be

$$\mu_e = \pi a_e^2 e/(\tau_e c); \quad \mu_p = \pi a_p^2 e/(\tau_p c)$$

and the angular momenta would be

$$m_e 2\pi a_e^2/\tau_e = h/2\pi; \quad m_p 2\pi a_p^2/\tau_p = h/2\pi$$

where $\tau_{e,p}$ are the orbital periods. (Here c is the speed of light and is required to get the right units for the magnetic moments.) Eliminating $\tau_{e,p}$ from these equations, it turns out that the radii also drop out and one finds

$$\mu_e = eh/4\pi m_e c; \quad \mu_p = eh/4\pi m_p c$$

These relations are correct in order of magnitude but because our current loop model oversimplifies the structure of both particles, each one must be multiplied

J. W. Halley, *How Likely is Extraterrestrial Life?*, SpringerBriefs in Astronomy, DOI: 10.1007/978-3-642-22754-7, © The Author(s) 2012

by a correction factor, called a g factor, in a more careful treatment. The order of magnitude of the energy of interaction of two magnetic moments is the produce of their moments divided by the cube of the distance betweent them

$$E \approx \mu_e \mu_p / a^3$$

This is roughly the difference between the energy of a hydrogen atom when the magnetic moments of the proton and the electron point in the same direction and when they point in opposite directions. Typically, hydrogen atoms in the interstellar medium will be oscillating between these two states and will be emitting and absorbing electromagnetic radiation with wavelength $\lambda = hc/E$ in the process. Combining the last three equations and putting in known values of the constants h, c, e, m_e, m_p gives radiation wavelengths of the order of a meter consistent with the more exact result of 21 cm.

Appendix 7.2
SETI Microwave Searches

Project	Phoenix
Band width	1200-3000MHz
Resolution	1Hz
Antenna diameter	305 m
Sensitivity	1×10^{-26} W/m^2
Number of stars searched	800
Project	SETI@home
Band width	1420 $\pm$ 1.25 MHz
Antenna diameter	305m (Arecibo)
Sensitivity	5×10^{-25} W/m^2
Number of stars searched	28% of sky 2 or 3 times
Project	SERENDIP IV
Band width	1420 $\pm$ 50 MHz
Resolution	0.6Hz
Antenna diameter	305m (Arecibo)
Sensitivity	5×10^{-24} W/m^2
Number of stars searched	28 % of sky
Project	SERENDIP III
Band width	424-436 MHz
Resolution	0.6Hz
Antenna diameter	305m (Arecibo)
Sensitivity	5×10^{-25} W/m^2
Number of stars searched	28 % of sky
Project	Argus
Band width	1420-1720 MHz

(continued)

J. W. Halley, *How Likely is Extraterrestrial Life?*, SpringerBriefs in Astronomy,
DOI: 10.1007/978-3-642-22754-7, © The Author(s) 2012

(continued)

Resolution	1Hz
Antenna diameter	100 3-5m dishes
Sensitivity	10^{-21} W/m^2
Number of stars searched	

Project	AUSTRALIA SOUTHERN SERENDIP
Band width	1420 $\pm$ 8.8 MHz
Resolution	0.6MHz
Antenna diameter	64m (Parkes)
Sensitivity	4×10^{-24} W/m^2
Number of stars searched	"southern sky"

Project	BETA
Band width	1400-1720 MHz
Resolution	0.6Hz
Antenna diameter	26m
Sensitivity	2.2×10^{-22} W/m^2
Number of stars searched	sky survey from -30° to 60° declination

Project	META III
Band width	1420.4-1557.3 MHz
Resolution	0.05Hz
Antenna diameter	30m Argentina
Sensitivity	1.0×10^{-30} to 7×10^{-25} W/m^2
Number of stars searched	sky survey of southern sky and 90 target stars

Appendix 7.3
FM and AM Signals

In human radio transmission, two methods, called amplitude modulation (AM) and frequency modulation (FM) are used to impose a signal on a 'carrier frequency'. The 'carrier frequency' f is the frequency to which you tune your dial, for example 91.1 MHz means a carrier frequency of 9.11×10^7 Hz or per second. If there is no signal on the carrier frequency then the electric and magnetic fields in the wave change with time according to the equation

$$carrier = A \times \sin(2\pi f t + \phi)$$

Here A is called the amplitude and ϕ is called the phase. The amplitude and phase will be different for the electric and the magnetic fields in the wave but we need not be concerned with that and we will set $\phi = 0$ for the carrier alone. If the signal is amplitude modulated, then the electric and magnetic fields vary as

$$AM\ signal = A(t) \times \sin(2\pi f t + \phi)$$

and if the signal is frequency modulated the fields vary as

$$FM = A \times \sin(2\pi f t + \phi(t))$$

Generally $A(t)$ or $\phi(t)$, which carry the information in the signal, vary much more slowly than the carrier wave does.

J. W. Halley, *How Likely is Extraterrestrial Life?*, SpringerBriefs in Astronomy, DOI: 10.1007/978-3-642-22754-7, © The Author(s) 2012

Appendix 7.4
Quantitative Analysis of Messages:
Correlations, Information, Entropy
and Complexity

If every symbol in a message is assigned a number a_α, then the 'pair correlation function' $\Gamma(d)$ where d is the number of symbols between a particular pair of symbols can be defined as

$$\Gamma(d) = \sum_{\alpha,\beta} a_\alpha a_\beta P_{\alpha,\beta}(d) - \left(\sum_\alpha a_\alpha P_\alpha\right)^2$$

α and β label the available characters. So for example you could choose $a_A = 1, a_B = 2, \ldots, P_{\alpha,\beta}(d)$ is the probability of finding α and β with d symbols between them in the message. P_α is the probability of finding α in anywhere in the message. This function can then be evaluated. A problem is that the choice of numbers a_α to associate with symbols α is arbitrary. In Ref. [3], Li tried various random assignments of numbers to letters to see if the correlation function was different for different assignments. The result, carried out for John Kennedy's presidential inaugural address in 1960 shows that the assignments do make a difference but the result is roughly the same for all of them and $\Gamma(d)$ behaves approximately as $d^{-0.9}$. The corresponding frequency spectrum would fall off very slowly with frequency, unlike the results of Voss and Clarke in Fig. 6 in Chap. 7 for music on the radio.

Another way to measure correlations in a message is to find an expression which measures the correlations between the information in one part of the message with the information in another part of the message. Suppose there are M symbols available. (For example using only the letters of the alphabet and one case would give M = 26.) If all possible messages are allowed, then in a message with N symbols the information is

$$\log_2 M^N$$

and the information per symbol is

$$\log_2 M$$

J. W. Halley, *How Likely is Extraterrestrial Life?*, SpringerBriefs in Astronomy, DOI: 10.1007/978-3-642-22754-7, © The Author(s) 2012

If all the symbols are equally likely, as they will be if all messages are allowed, then the probability P_α of any symbol is $P_\alpha = 1/M$ so that the expression for the information per symbol is

$$-\log_2(1/M) = (1/M) \sum_\alpha \log_2 P_\alpha = -\sum_\alpha P_\alpha \log_2 P_\alpha$$

But now suppose there are constraints on the use of the symbols so that all the symbols are not equally likely to be used. Suppose that on average, in a message of N symbols, the number of symbols of type α is N_α. We imagine all the symbols in the message with N symbols written out. Now we scramble them in all possible ways. The total number of ways to scramble them is $1 \cdot 2 \cdot 3 \cdots N = N!$. But all of those scramblings (called permutations) which only rearrange identical symbols do not give a new message. As a result, the total number of messages in this case is

$$\frac{N!}{N_1!N_2!\ldots N_M!}$$

To put this in a convenient form, one has to use the mathematical fact that, if a number L is large, then to a good approxiation

$$\log_2 L! \approx L \log_2 L - L/\ln 2$$

Using this, the information per symbol in a message in this constrained case can be shown using a little algebra to be

$$(1/N)\log_2 \left[\frac{N!}{N_1!N_2!\ldots N_M!} \right] \approx -\sum_\alpha P_\alpha \log_2 P_\alpha$$

in which $P_\alpha = N_\alpha/N$. This has mathematically the same form as I found above for the information in the case that all the symbols were equally likely to be used.

So far, this does not involve correlations. But now consider two symbols, of types α and β respectively, with n symbols between them in the message. Count the number of such pairs in a message of length N. (For example, count the number of pairs A–B with one intervening symbol in the case n = 1.) Call the number of such pairs $N_{\alpha\beta}(n)$ Averaging over all messages, the probability of this combination is $P_{\alpha\beta}(n) = N_{\alpha\beta}(n)/N(n)$ where $N(n)$ is the total number of pairs separated by n symbols in the message. Now we can find the part of the information in the message which is contained in the pairs separated by n symbols by following an argument identical in form to the one in the preceding paragraph. Scramble the pairs in all possible ways and take account of the fact interchanging two identical pairs gives nothing new, so the number of ways to arrange these pairs is

$$\frac{N(n)!}{N_{11}(n)!N_{12}(n)!\ldots N_{MM}(n)!}$$

Taking the $\log_2$, the information per pair is

$$-\sum_{\alpha,\beta} P_{\alpha\beta}(n) \log_2 P_{\alpha\beta}(n)$$

If the constraints on message formation are such that a symbol α at one point in the message has no effect whatever on the probabiity of finding a symbol β n symbols away, then the probability $P_{\alpha\beta}(n) = N_\alpha N_\beta/N^2 = P_\alpha P_\beta$. This is the case of no correlations. If there are no correlations, then we expect the information in the pairs to be larger than in the case in which the presence of the symbol α does affect the likelihood of finding the β n symbols away. (For example u always follows q in English and the u after the q obviously conveys no new information.) To isolate this effect in a mathematical function Shannon defined the information correlation function $C_{\alpha\beta}(n)$ to be the difference between the pair information and the value which the pair information would have if there were no correlations:

$$C_{\alpha\beta}(n) \equiv -\sum_{\alpha,\beta} P_{\alpha\beta}(n) \log_2 P_{\alpha\beta}(n) + \sum_{\alpha,\beta} P_\alpha P_\beta(n) \log_2 P_\alpha P_\beta(n)$$

$$= -\sum_{\alpha,\beta} P_{\alpha\beta}(n) \log_2 P_{\alpha\beta}(n) + 2\sum_\alpha P_\alpha \log_2 P_\alpha$$

From the discussion, this should be negative or zero. This is the quantity which Li evaluated in Ref. [3] using Shakespearean plays, the bible in German and other human texts. He found that it was characteristically getting smaller with increasing n approximately as $1/n^3$. (Li plots the negative of $C_{\alpha\beta}(n)$ in Ref. [3].) However there were definitely finite correlations, demonstrating explicitly that human language does not maximize information content.

References

1. B. Carter, Phil. Trans. R. Soc. A **310**, 347 (1983)
2. R.J.M. Fry, J.T. Lett, Nature **335**, 365 (1988)
3. W. Li, Complex Syst. **5**, 381 (1991)

Index

1/f noise, 89

A
Active SETI, 88
Adenine, 32
Allen telescope, 84
Amino acids, 33
Anthropic principle, 40, 49, 52, 114, 116–117
Antibiotic resistant bacteria, 53
Anthropogenic electromagnetic signals, 89
Arecibo, 84, 86
Astrobiology, 76, 95
Autocatalysis, 43
Autocatalytic cycles, 39, 43

B
Band width, 85
Big Bang, 11, 12
Biochemistry, 3, 6, 23, 32
Black body temperature, 26

C
Cairns-Smith, 39
Callisto, 101
Calvin cycle, 45
Cassini, 102, 104
Catalysis, 39
Catalysts, 33
Cell walls, 33
Civilization lifetimes, 69, 73
Colonization, 4, 66
Complexity, 5–6, 32, 46, 90
Conspiracy, 63–64
Contemplation hypothesis, 64

Correlations, 89, 90, 93, 135
Crab nebula, 14
Cytosine, 32

D
Darwin, 38, 47, 53
Deoxyribonucleic acids (DNA), 32
Diffusion, 70
Doppler shift, 11, 20, 123
Drake equation, 3–6, 15, 17–19, 31, 107, 121
Dyson, 42–43, 47
Dyson spheres, 83

E
E. Coli, 33, 36
Eigen, 39, 42
Endosymbiosis, 47
Entropy, 47
Environmental ethics, 110
Enzymes, 33
Eukaryotes, 47, 54
Europa, 95, 100–102
Evolution, 12, 15–16, 27, 31–33, 38–39, 41–42, 44, 46–47, 50, 52, 54
Evolutionary paradox, 38–39
Exobiologists, 96
Extinction probabilities, 74
Extremophile, 76
Extremophiles, 102

F
Falsifiability, 59, 61
Fermi paradox, 41, 51, 59
Filling time, 71–73

J. W. Halley, *How Likely is Extraterrestrial Life?*, SpringerBriefs in Astronomy, 149
DOI: 10.1007/978-3-642-22754-7, © The Author(s) 2012